D0480341

The Indisputable Existence of Santa Claus

THE MATHEMATICS OF CHRISTMAS

Dr Hannah Fry
&
Dr Thomas Oléron Evans

BLACK SWAN

TRANSWORLD PUBLISHERS
61–63 Uxbridge Road, London W5 5SA
www.penguin.co.uk

Transworld is part of the Penguin Random House group of companies whose addresses can be found at global.penguinrandomhouse.com

First published in Great Britain in 2016 by Doubleday
an imprint of Transworld Publishers

This edition published in Great Britain by Black Swan
an imprint of Transworld Publishers

A CIP catalogue record for this book
is available from the British Library.

ISBN 9781784162740

Typeset in 10/14pt Charter by Julia Lloyd Design
Printed and bound by CPI Group (UK) Ltd, Croydon, CR0 4YY

Penguin Random House is committed to a sustainable future for our business, our readers and our planet. This book is made from Forest Stewardship Council® certified paper.

1 3 5 7 9 10 8 6 4 2

CONTENTS

INTRODUCTION

I wish it could be Christmas every day

The nights are drawing in, the crisp winter air is filled with the inviting smells of roasted chestnuts and mulled wine, and children everywhere are struggling to contain their excitement about Santa's imminent arrival. It can only be Christmas – the most magical time of the year.

True, your overdraft is straining under the weight of your generosity, you're obliged to spend hours writing heartfelt festive wishes to close friends that you've accidentally forgotten about for the rest of the year and your daily alcohol consumption barely ever dips below the recommended weekly limit, but you wouldn't have it any other way.

With so much goodwill everywhere, it's no wonder glam-rockers Wizzard wished it could be Christmas every day.

But let's ponder that for a moment and run the numbers. What would it be like if we decided, as a nation, to move towards a System for Christmas Repeating Every day At Midnight (SCREAM for short)?

One big winner of SCREAM would be the UK's largest tinsel factory, Festive Productions in Cwmbran, south Wales, who would see a substantial increase in turnover. Since we Brits normally keep our decorations up for around a month, we'd need to replenish our tinsel supplies 12 times as frequently as we do now. That means Festive Productions, who currently make 600,000 metres of tinsel a day,* would have to up their daily output to a whopping 7.2 million metres.

Of course, all that tinsel would be useless without a steady supply of Christmas trees to dangle it on. Currently, the UK buys 8 million trees a year, compared to 40 million in the USA and 42 million in the rest of Europe.† Since each tree needs seven years to grow from a sapling before it's worthy of becoming the dazzling centrepiece of your family home, switching to SCREAM could result in a shortfall that would take some years to rectify.

However, given that there are 350 million fir trees being grown on Christmas tree farms in North America alone, if we ration ourselves to one new tree a month we'll be OK until at least May or so.

After that, we can utilize some of the other 3.04 trillion trees around the world‡ while we're waiting for the fir farms to replenish. True, it might not have that pine-fresh smell, but if you hack up a willow and slap a bit of tinsel on it, it should do for a few years. But choose your tree wisely. A 120-metre redwood poking out of your living-room window probably

* Source: Festive.co.uk.

† H. Preston, 'For a Very Merry Christmas, Invest in Trees for the Season', *International Herald Tribune*, 23 December 2000.

‡ T. W. Crowther et al., 'Mapping tree density at a global scale', *Nature*, 2015, vol. 525, no. 7568, pp. 201–5.

won't go down too well with the council.

The daily festive feast could cause even more upheaval. Each year the British public buys and eats 10 million turkeys for our Christmas dinners. From hatching to table, turkeys live for 6 months, so around 1.8 billion turkeys would need rearing and housing at any one time to meet the SCREAM demand.

The University of New Hampshire* suggests each turkey should have at least 6 ft^2 of covered shelter. That translates into a giant turkey coop of 1,003 km^2, around two-thirds the size of Greater London.

As Britain is a nation of animal lovers we're certain most supporters of SCREAM would insist on all turkeys being free-range, which means we'd also need to factor in 100 ft^2 of pasture for each bird. This translates into a colossal turkey farm of 17,726 km^2 that would cover all of London and the home counties and no doubt be the envy of the world.

Sure, we'd be losing our nation's capital, financial institutions, parliament and 2,000 years of history, but think how much joy and Christmas cheer we'd stand to gain. Overall, it's a small sacrifice to make.

The benefits don't stop there either. Those turkeys are going to generate a lot of poo: 4 billion kg a year, in fact.† As you might expect, this brown gold serves

* D. C. Seavey and J. C. Porter, 'Housing and Space Guidelines for Livestock', *University of New Hampshire Cooperative Extension*, 2009.

† G. T. Tabler et al., 'Poultry Litter Production and Associated Challenges', *University of Arkansas, Cooperative Extension Service*, 2009. Litter of 2.3 lb per bird equates to just over 1 kg of poo for each turkey in its lifetime. Hence 10 million new birds a day generate approximately 4 billion kg a year.

as an excellent fertilizer, but more impressively it can also be used to fuel power stations. Under SCREAM, then, all our energy concerns will be sorted and the flowers in Kew Gardens will flourish (though the resident turkeys may not appreciate them as much as the human visitors once did).

If we decide not to give up our entire south-east, we could scatter the turkey farms around the country, using buildings we no longer have a need for, like schools, stadiums and shopping centres.

We'll want to keep the hospitals, though. If we're all eating the Christmas average of 6,000 calories a day, 3,500 calories more than any of us need, we'll each have put on around 6 st 12 lb by the end of March.* The increased prevalence of obesity might make our hospitals busy, but it's great news for the manufacturers of reinforced wider beds and elasticated trousers.

SCREAM won't necessarily be stress-free for everyone. The *EastEnders* scriptwriters will have to concoct a brand-new dramatic hour-long special every day, and Santa may well go into meltdown after a matter of months. But think how great it will be for the children who wake up to find a stocking full of presents every single morning.

Granted, our GDP might slip a bit if almost everyone stopped going to work. There's also the small matter

* M. Wishnofsky, 'Caloric Equivalents of Gained or Lost Weight', *American Journal of Clinical Nutrition*, 1958, vol. 6, no. 5, pp. 542–6, states that 1 lb of body fat contains about 3,500 calories.

of having nowhere to shop for the presents, 1.8 billion turkeys running amok and a generation of children growing up with no work ethic or sense of discipline. But let's not get bogged down by trivialities like that.

On balance, we can see hardly any convincing arguments against SCREAM. Unfortunately, the 'experts' of the nation are not to be persuaded. We're stuck with Christmas as an annual event, and we're going to have to make the best of it.

So, to really extract every possible ounce of fun and joy from the day, there's only one sensible way to plan your festivities: using mathematics.

How else but with mathematics could you work out how to time cooking your Christmas turkey to perfection? Or determine the best way to wrap your presents? Or design a flawless Secret Santa system? Or guarantee you beat your family in the annual board-game argument?

Some people might try to accuse us of taking the magic out of Christmas by reducing it to a set of mathematical equations, but for us the exact opposite is true.

We believe that mathematics is so powerful that it has the potential to offer a new way of looking at anything – even something as warm and wonderful as Christmas. Maths can uncover hidden patterns behind the familiar festivities and provide unique insights into how to really get the most out of your traditional celebrations. All of which, we think, adds up to make this time of year even more magical.

We hope to persuade you by showing you how to use mathematics to dress your tree with flawless precision. We'll give you statistics to predict what the Queen is going to say in her Christmas message. We'll even offer ultimate proof, if ever it were needed, of Santa Claus's existence.

So curl up in front of the fire, pour yourself a large warm glass of mulled wine, pop on a CD of Cliff Richard's greatest Christmas hits, and enjoy the merriest mathematics of Christmas.

CHAPTER 1

The indisputable existence of Santa Claus

It is astonishing that some people still doubt the existence of Santa Claus. Despite the vast amount of photographic evidence, the hundreds of annual reports on Father Christmas's activities from perfectly reputable news sources and the bulging stockings full of presents that reliably appear on Christmas morning, somehow the doubters remain unconvinced.

Thankfully mathematics can help.

The conspiracy theorists have already tried turning to science to demonstrate their (clearly incorrect) position. They calculate that if Santa were to visit the 1.9 billion children in the world,[*] he would have to travel at 3,000 times the speed of sound while carrying around 300,000 tonnes of presents[†] (about the weight of six *Titanics*). Richard Dawkins, king of the sceptics, has insisted that the lack of any noticeable sonic booms from all that zipping about at supersonic speeds is more than enough evidence that Santa cannot possibly be real.[‡]

Worse still, some claim that this astonishing

* Source: UN population data from 2015, 0–14 yrs.
† H. Romesburg, *Best Research Practices* (2009), p. 190.
‡ R. Dawkins, *Unweaving the Rainbow* (2006), p. 141.

weight of parcels travelling at such a remarkable speed would practically vaporize the leading reindeer, who would have to withstand the full hit of air resistance. Meanwhile, sitting in the back of his sleigh, Santa would be subjected to forces tens of thousands of times stronger than gravity, making it impossible for him to breathe or to retain any of the physical structure of his bones or internal organs, thus reducing him to a liquefied mess. While this would admittedly explain how he was able to slip down some of the narrower chimneys on his route, it probably wouldn't make for very attractive Christmas cards.

Sure, all these scientific spoilsports *sound* convincing enough. Although their arguments totally depend on the assumption that Santa isn't a macroscopic quantum object capable of being in two places at once. And that he's unable to manipulate time (though how else do they think he manages not to age in photographs?). *And* that he hasn't constructed a NASA-style heat shield to protect his reindeer. Or invented a device to suppress sonic booms.

They also assume that any of these simple explanations is *more* far-fetched than the idea that the vast majority of the adult world is participating in a massive conspiracy, with parents cheerily lying to their children on behalf of a mystical non-existent figure, postal services fiendishly filtering out letters to Santa rather than returning them to sender as they normally would, and news agencies annually publishing blatant falsehoods that go against all their journalistic ethics, simply to maintain the whole pointless charade. Riiiggght. Sure.

The sceptical scientists' arguments also illustrate an important point. The great difference between scientific and mathematical proof.

The scientific method takes a theory – in our case that Santa is real – and sets about trying to prove that it is false. Although this may seem a little counterintuitive on the surface, it actually does make a lot of sense. If you go out looking for evidence that Santa doesn't exist and don't find any . . . well then, that is pretty revealing. The harder you try, and fail, to show that Santa cannot exist, the more support you have for your theory that he must. Eventually, when enough evidence has been gathered that all points in the same direction, your original theory is accepted as fact.

Mathematical proof is different. In mathematics, proving something 'beyond all reasonable doubt' isn't good enough. You have to prove it beyond all unreasonable doubt as well. Mathematicians aren't happy unless they have demonstrated the truth of a theory absolutely, irrefutably, irrevocably, categorically, indubitably, unequivocally and indisputably. In mathematics, proof really means proof, and once something is mathematically true, it is true for ever. Unlike, say, the theory of gravity – hey, Newton?

If we want to silence the doubters once and for all we have to turn to mathematical proof.*

* This may or may not involve being slightly cheeky with some mathematical rules along the way.

So here we go then. Let's use some mathematical logic to see if we can prove the indisputable existence of Santa Claus, starting with the statement on the opposite page . . .

Everything
on this page
&
everything on page
152
is false . . .

Back with us? Good.

Those two statements – the one on page 11 and the one on page 152 – are all we need to prove the existence of Santa.

If the statement on page 11 is true in its claim that everything on the two pages is false, then Santa doesn't exist, the sceptics win and we're not looking good for Christmas this year.

But if everything on those two pages is false, then the statement on page 11 must itself be false. But that's a contradiction, because we just supposed that it was true.

That means the statement on page 11 can't possibly be true. And if it is not the case that 'Everything on page 11 and everything on page 152 is false', then at least one of the two statements must be true.

But hang on, we've already decided that the statement on the previous page can't be true because it contains a contradiction, so the true statement must be the one on page 152. Hence Santa exists![1]

Not convinced? OK, how about a different approach.

Who's to say there's only one Santa, after all? Let's try and determine if 'Santas' are real. Some sort of secret society of Santas, perhaps, passing their festive baton from generation to generation. We don't really care about the dead ones, of course, only the ones that exist now. And in terms of existing Santas, there are only two possible statements that can be made:

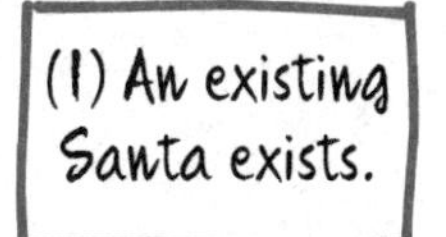

(2) An existing
Santa does not exist.

We can all agree that one of those must be true, right?

But hold on, statement (2) is contradictory. How can an existing Santa not exist? A 'Jolly Santa' must be jolly by definition, a 'Portly Santa' must be a bit on the chubby side and so an 'Existing Santa' *has* to exist.

That means that statement (2) must be false – but since we agreed that *one* of the two statements was true, the only logical conclusion is that (1) is true and so at least one Santa must exist.[2]

Still need more?

All right, one more go.

We can surely all agree that at least one of *these* two statements is true:

(1)
Santa exists.

(2)
1 + 1 = 2.

Well let's play with the second of those a little.

Let's put $a = 1, b = 1$.
Then:

$$a = b$$

$$a^2 = b^2 \quad \text{squaring both sides}$$

$$a^2 - b^2 = 0 \quad \text{taking } b^2 \text{ away from both sides}$$

$$(a - b)(a + b) = 0 \quad \text{factorizing* the left-hand side}$$

$$\frac{(a - b)(a + b)}{(a - b)} = \frac{0}{(a - b)} \quad \text{dividing by } (a - b)$$

$$\frac{\cancel{(a - b)}(a + b)}{\cancel{(a - b)}} = 0 \quad \text{cancelling the } (a - b) \text{ on the left-hand side}$$

$$a + b = 0 \quad \text{tidying everything up}$$

$$1 + 1 = 0 \quad \text{putting } a = 1 \text{ and } b = 1 \text{ back into the left-hand side}$$

Whoa there a minute! $1 + 1 = 0$?!

Well, if $1 + 1 = 0$ and not 2 as we thought, then statement (II) is false.

But if we agreed that at least one of the two

* Factorizing is a way of turning a sum into a product by breaking an expression down into its factors. It's easier to see that $(a - b)(a + b) = a^2 - b^2$ if you work backwards. Take $(a - b)(a + b)$ and multiply the two brackets together to get $(aa + ab - ba - bb)$. The two terms in the middle cancel to leave $a^2 - b^2$.

statements was true – and we did, there's no backing out of it now – then if it isn't (II), it must be (I). Hence Santa exists.[3]

So there we have it. Not one, not two, but three mathematical proofs of the existence of Santa Claus. The issue has clearly been resolved once and for all. Come Christmas Eve, you can tuck yourself up in bed, safe in the knowledge that a fat man in red with a handwritten list of all the bad things you've ever done will definitely be squeezing his way down your chimney as you sleep. Our work is done.

. . .

Seriously, there's nothing to see here. You can skip forward to the next chapter now.

. . .

Oh, all right, all right! There may have been the tiniest bit of logical trickery going on in those proofs you saw earlier. If you really are absolutely determined to know the truth, there are some notes at the end of this chapter that should shed a little more light on them, but let's not lose sight of the big picture here.

While the proofs in this chapter may be a bit iffy (and seriously, we're trying to prove that Santa exists here, cut us a little slack), the *concept* of proof is the surest thing of all. Taking your existing knowledge and building new truths using nothing but logical deduction, knowing that every fact you learn can be rigorously demonstrated from everything that has gone before . . . that is what makes maths so powerful. For many people, it also makes it beautiful.

If you're thinking that coming clean over the slightly mischievous nature of our proofs was an admission that the big guy *doesn't* exist, well then, think again. Because, dubious as you may be about these proofs that Santa is real, it is doubtful that anyone could come up with a convincing mathematical proof that he *isn't* real either.

It turns out that in mathematics, no matter how firmly you lay your logical foundations, there are always some things you just can't prove one way or the other. These special statements are called 'undecidable'. You can write them down and understand them, they are either true or false, but you can never discover which. You just have to take a deep breath and decide to accept them or not.

The existence of these logical black holes was discovered by a mathematician called Kurt Gödel in the 1930s, and they have fascinated mathematicians ever since. All this doesn't mean maths is wrong, of course; these slippery statements are far removed from the maths that is used in everyday life and 2 + 2 won't be equal to 5 any time soon, but they do suggest an intriguing new perspective on our bearded friend.

Because maybe that's how we should see Santa. An undecidable being. You can't prove his existence one way or the other, you just have to close your eyes and decide for yourself.

And deep down, we all know the truth, don't we?

ENDNOTES

[1] This is a version of the Liar Paradox. The classic formulation of the paradox is generally along the lines of: 'This statement is false' or 'Everything I say is a lie'.

These statements can't be true because in being so they would also necessarily be false. Likewise, they can't be false because that in turn would make them true.

In our version, including the statement 'Santa Claus exists' means we can get away without a paradox, though only if that additional statement is true. We could have put anything we wanted on page 152 and proved all manner of things: 'Unicorns are real', 'Mathematicians are the best dancers', 'This book will self-destruct in five seconds'. . . anything at all.

Mathematically speaking, all this is a bit naughty because the statement 'Everything on this page and everything on page 152 is false' refers to itself, just like 'Everything I say is a lie' in the original form of the paradox. Self-referential statements like these don't actually have to be true *or* false, which resolves the paradox.

[2] This proof is similar to Descartes' ontological proof of God's existence. His version of the proof went something like this:

- By definition, God is a being that is perfect in every way.
- A being certainly could not be perfect if it did not exist.

- Therefore God exists.

Forms of this proof have been debated by philosophers for centuries, so we're certainly not going to be able to do that debate justice here, though this does illustrate how closely aligned mathematical logic is with philosophy.

Speaking loosely, one problem with this proof is that the lack of existence of God is only problematic if you have *already* assumed that God exists. Similarly, while it is true that *if* an 'existing Santa' exists, then it must exist, that's not really saying much. If an 'existing Santa' *doesn't* exist then there is nothing to cause a contradiction.

[3] Of course, this 'proof' that $1 + 1 = 0$ has a flaw in it. It comes in when dividing by $(a - b)$. Since $a = b$, you are effectively dividing by zero in this step, something which you'll no doubt remember your maths teachers telling you is a very bad idea. Whenever a zero sneaks into the bottom of a fraction the rules of arithmetic, which are normally well behaved, break down, meaning you end up with all sorts of infinities and contradictions. No one has yet come up with a good suggestion for what the result of dividing by zero should be and so mathematicians say it is *undefined*. Anything in the proof beyond that point can't be trusted since it is built upon a logical flaw.

If you're interested in the real proof that 1 + 1 does indeed equal 2, you can find it in Alfred North Whitehead and Bertrand Russell's *Principia Mathematica* (1910). It takes them 362 pages to get there, and even then, it only applies 'when arithmetic addition has been defined'.

CHAPTER 2

Decorating the tree

By the time you get to the end of this book you'll have all the tools you need to plan the perfect mathematical Christmas. We'll come on to the important matters of presents, food and games a little later. But first things first – before your Christmas planning gets under way in earnest, you'll want to get off on the right foot with a geometrically superior Christmas tree.

In some senses, designing the perfect tree is simple. All you need is a neat, symmetrical placement of tinsel and lights, accented by an equally precise arrangement of spectacular twinkling baubles.

But how long does your tinsel or string of fairy lights need to be to wrap around your tree perfectly? Rather than relying on guesswork, there is a simple equation you can use to help you.

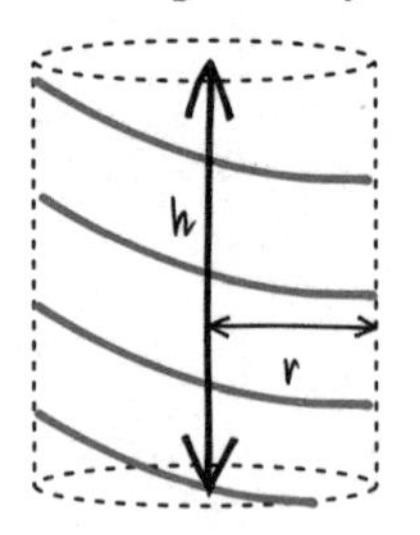

To give you an idea of all the festive glory* you can hope for with this calculation, we've included a sketch of your tree on the left.

You did buy a cylindrical tree, right? Even if you didn't, this version of the formula works for particularly fat

* We're mathematicians, not artists.

Christmas trees that are approximately cylinders, save for the spindly bit at the top which you don't cover in tinsel anyway.

In red is the tinsel wrapped around the tree. The height of the cylindrical section is h and the radius is r.

If you unravelled the cylinder to look at only the surface of your Christmas tree, it would look something like the following: a rectangle of height h and width equal to the circumference of your tree, $2\pi r$.

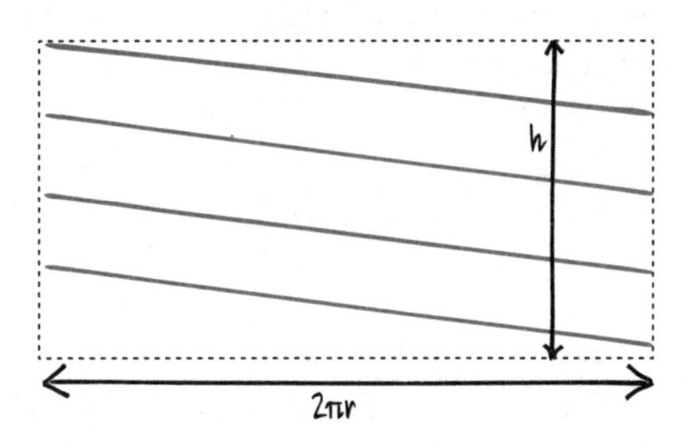

The diagonal lines represent loops of tinsel around your tree. As each line disappears off the edge on the right-hand side, it appears as the beginning of the next diagonal line on the left – like a pine needle Pac-Man.

Exactly how many loops your tree should have we'll leave to your artistic talents (we exhausted ours on that sketch above), but once you've made your decision (let's say you have n loops), the length of the tinsel can be easily calculated by zooming in on one of these diagonal lines.

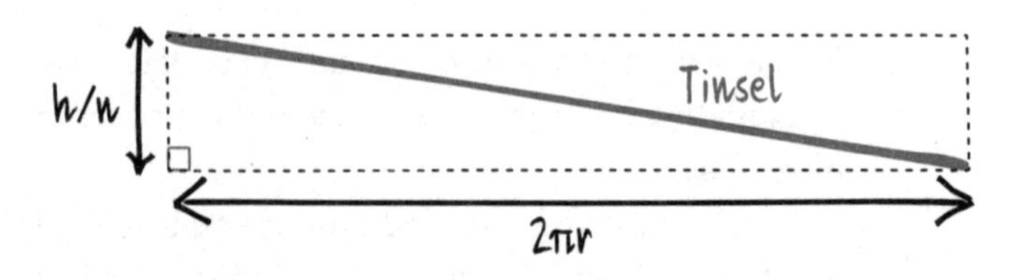

Since you want everything to be evenly spaced, the vertical distance each loop must cover is h/n. The width is just the circumference of your tree, $2\pi r$, and, since the length l of the tinsel on this particular loop forms the hypotenuse of a right-angled triangle, Pythagoras does the rest for us:

$$l = \sqrt{\left(\left(\frac{h}{n}\right)^2 + (2\pi r)^2\right)}$$

Who said Pythagoras would never be useful?

Taking all n loops into account, the total amount of tinsel required, L, is given by the following formula.

The Length of Tinsel (L) Needed to Decorate a (cylindrical) Christmas Tree

$$L = n\sqrt{\left(\left(\frac{h}{n}\right)^2 + (2\pi r)^2\right)}$$

h is the height of your tree
r is the radius of the base
n is the number of loops

'But hang on!' we hear you cry. 'My tree isn't cylindrical!' If anything, it looks more like a cone. Typical mathematicians making sweeping unjustifiable assumptions . . .

Well, hold on to your accusations for now.* While you can use the cylindrical version as an upper limit on your required amount of tinsel, even if your tree gets a lot skinnier at the top, we've also done the calculation for a tree that takes the more traditional shape of a cone.

The maths isn't for the faint-hearted, but if you're keen not to waste tinsel you can find the equation in the notes at the end of the chapter.[1]

For the rest of us, happy with our geometrically dubious trees, we can move on to the matter of baubles.

Spherical baubles seem to be the traditional ornaments with which to adorn a tree and in any other setting we're big fans of spheres. But while they may catch the light rather beautifully, to a mathematician they're just not very . . . *festive.*

If we're being honest with ourselves, we can all agree that there is nothing more Christmassy than corners.

Not just any corners, of course. We want neat, symmetrical, mathematically superior corners. Which is why, for our tree decorations, we're turning to our five favourite shapes; objects that have been admired and studied by geometers for thousands of years.

* There'll be plenty of room to throw that particular insult at us later in the book.

Ladies and gentlemen, may I introduce you to . . .
The platonic solids:

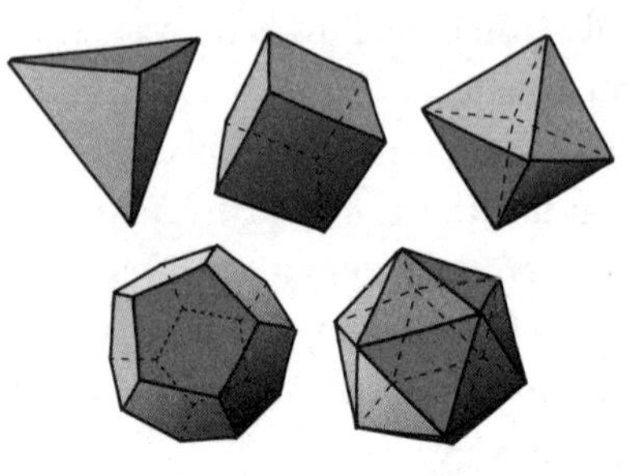

The platonic solids are 3D shapes where every face is a regular polygon. Around a face, every side is the same length and every angle is the same size. Looking at each shape as a whole, every face is identical and every corner is the same as every other.

Our platonic solids are: (top row, left to right) the tetrahedron, the hexahedron (which often goes by its street name of 'cube'), the octahedron, (second row) the dodecahedron and the icosahedron.

These five beauties are the only three-dimensional shapes that fit our criteria. We know this, not because we've checked all possible three-dimensional shapes, but because we can prove it. To find out how, see the notes at the end of the chapter.[2]

Any of these five platonic solids would look marvellous on a mathematical Christmas tree. After careful consideration, we reckon the hexahedron offers the perfect aesthetic balance between simplicity and cornery goodness. You can make your cubic baubles using the following template.

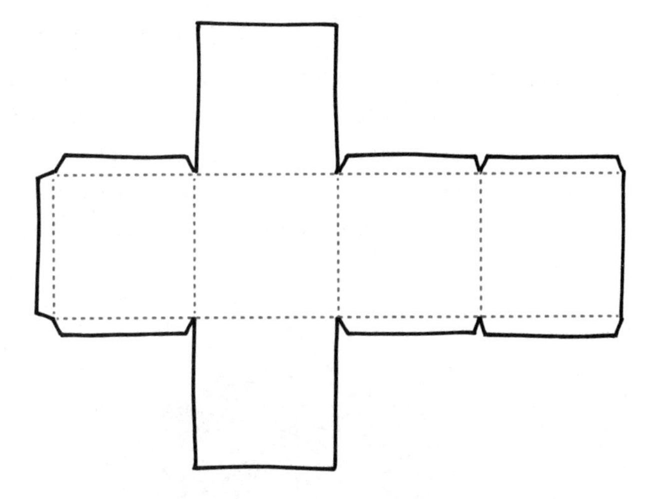

With your tree bedecked with fantastically festive Christmas cubes, you're sure to be the talk of the town.

. . .

Hmm . . . For some reason, our editor seems to think that cubical baubles are insufficiently ambitious. Well then, in case you also can't appreciate the supreme elegance and mathematical purity of the cube, we've put together some instructions for making your very own stellated icosahedral baubles.*

The icosahedron is the basis for this shape. Technically, a stellation is where you extend out the

* Those who'd like to stick to the original platonic solids for their Christmas ornaments can find templates for a tetrahedron and an octahedron here: (senteacher.org/worksheet/12/NetsPolyhedra.html). James Grime has also put an excellent instructional video on how to make a Post-it-note dodecahedron on his YouTube channel: *singingbanana*.

planes of each face until they meet beyond the boundary of the original shape. In this case, it's just like adding a point to each of the faces of your icosahedron.

The end result here is a 20-pointed three-dimensional star that will look superb as a sparkling bauble or, when supersized, perfect taking pride of place at the top of your tree.

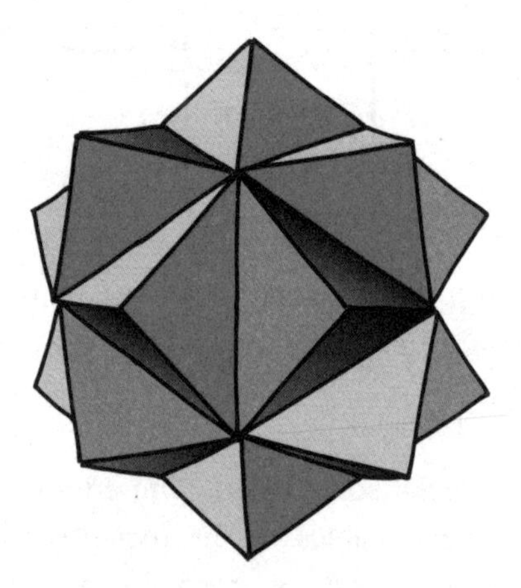

Here's how to make it.*

This origami-style technique involves making a series of *modules* from smaller pieces of paper and then slotting them together to create the final shape.

You need to start off with a square piece of paper. Some people suggest using Post-it notes, others recommend tinfoil, but for that extra Christmassy glitz, we think metallic wrapping paper is the way to go.

* Our instructions are adapted from a video by Sherri Burroughs: youtube.com/watch?v=nL2mvtdhiq8.

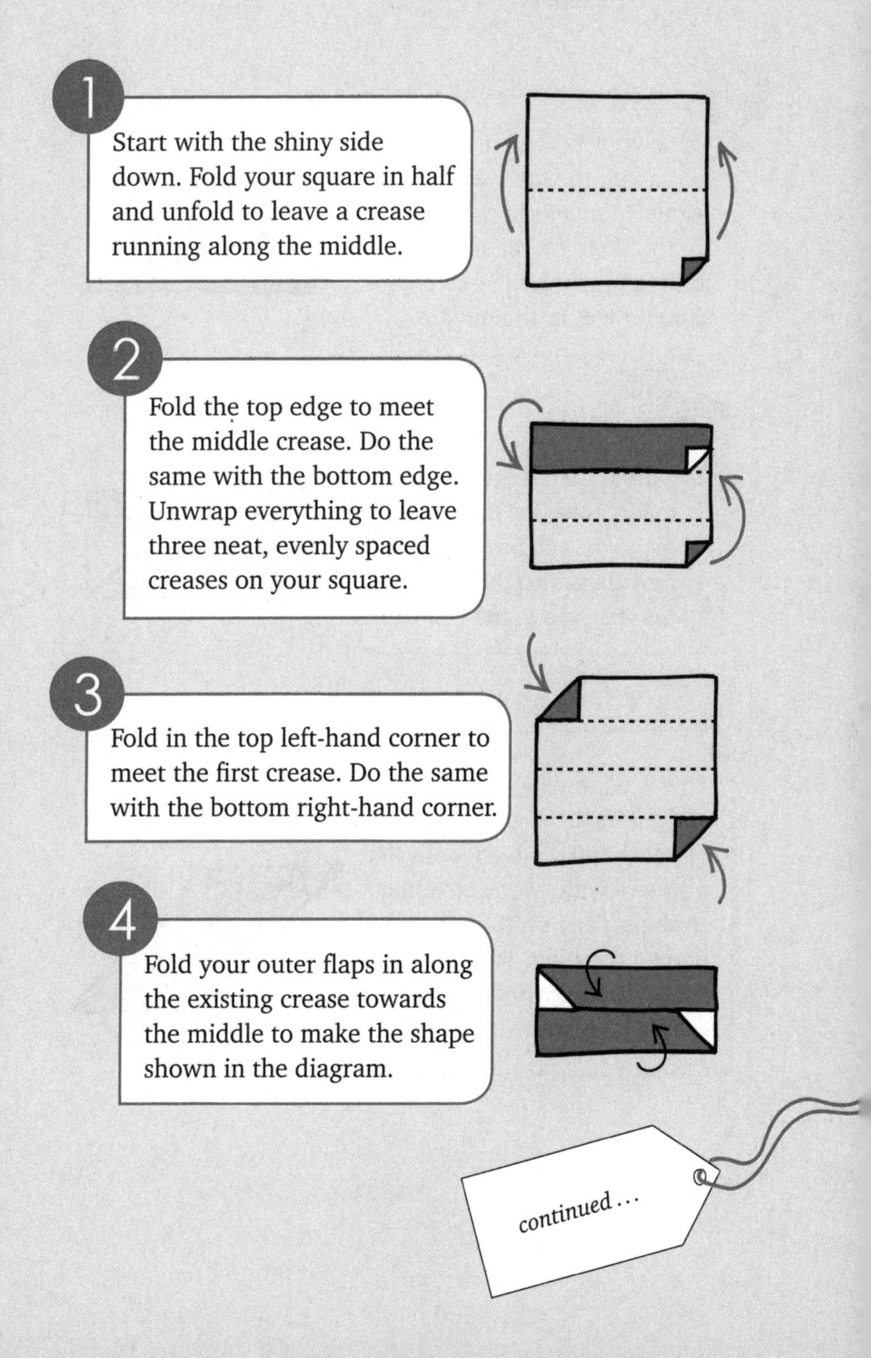
1
Start with the shiny side down. Fold your square in half and unfold to leave a crease running along the middle.
2
Fold the top edge to meet the middle crease. Do the same with the bottom edge. Unwrap everything to leave three neat, evenly spaced creases on your square.
3
Fold in the top left-hand corner to meet the first crease. Do the same with the bottom right-hand corner.
4
Fold your outer flaps in along the existing crease towards the middle to make the shape shown in the diagram.
continued …

5

Using your bent corners as a guide, fold down the top right-hand corner to meet the outer edge. Do the same on the left-hand side, folding upwards. The two flaps should meet in the middle.

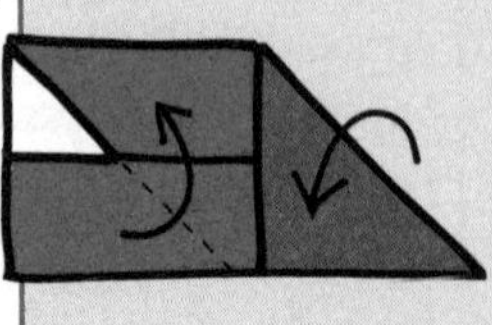

6

Unfold the creases you just made. Now for the tricky bit – you need to tuck these new flaps into the pockets that run along the centre.

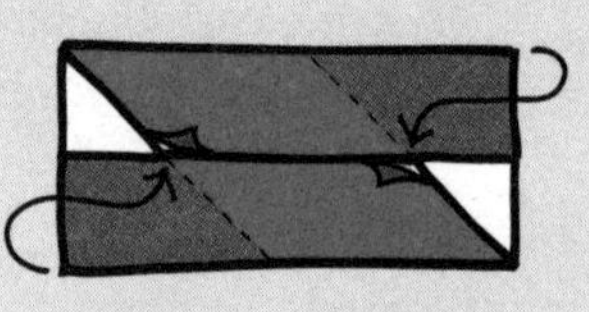

7

This is what you should end up with: a tucked cross in the centre of your shape, which looks like a square with two protruding triangles, one on the left and one on the right. Fold these two triangles backwards along the crease lines shown.

8

Finally, crease your entire shape by folding towards you along the vertical line shown.

Unwrap steps 7 and 8. Once you unfold your shape to reveal the creases, it should look something like the diagram in step 9 below. One side of the module will be smooth, the other, facing upwards, will have a cross on top made from tucked paper folds.

9

Repeat 29 times to give you 30 modules in total.

If this step has punctured your enthusiasm somewhat, cheer yourself through the pain barrier with thoughts of our tormented editor surrounded by bits of paper cursing the day she ever spoke badly about the cube.

10

Once you've made 30 of the modules, lay them out on a table with the crosses on top.

Interlock the modules by tucking the protruding triangles of each piece into the cross-hatched pockets of their neighbours as shown in the image above. They should slide in neatly, and the folds should all line up.

Carry on slotting modules together in the same way, securing with glue if necessary, until all 30 are in place. The final few will be a little tricky, but the shape of the icosahedron should emerge naturally from the creases in the paper. If all goes to plan, you should end up with something that looks like the image on page 26.

And there you have it. Simply repeat this entire process 20 or so times to create 600 modules which will slot together for a complete set of baubles, plus 30 more with larger squares of paper for the star on the top of your tree. And, assuming it's still December, you'll have a festive centrepiece bedecked in mathematically glorious decorations that will be the envy of all of your friends at your Christmas party.

At which point you can just sit back, add some ice regular hexahedrons to your rum punch and marvel at your creation.

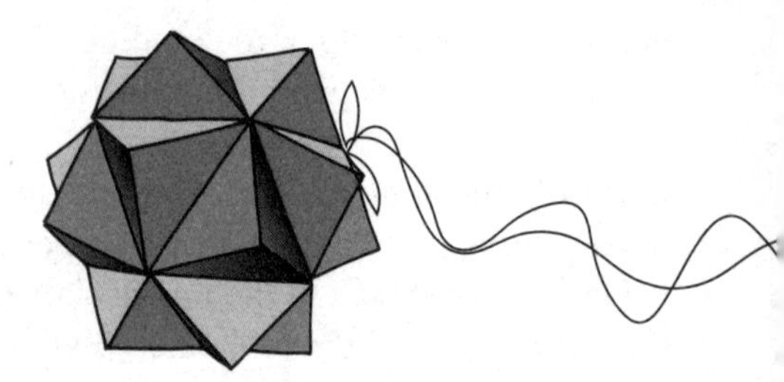

ENDNOTES

[1] Having a cone for a Christmas tree makes the maths a lot trickier. Each loop of the tinsel or fairy lights will now need to lie diagonally along a section of an *annulus* (what you're left with once you've chopped your flattened cone up into *n* parts) so using triangles is out.

Instead, we can consider an Archimedes spiral along the surface of the cone. If *h* is the height of your tree, *r* is the radius of the base and *n* is the number of loops you want the tinsel to make, the spiral will dictate exactly where the tinsel must sit as you go up the tree (*z*).

The formula for the position of your tinsel as you move up the tree is given by the following:

$$x = \frac{(h - z)}{h} r \cos \frac{2\pi n z}{h}$$

$$y = \frac{(h - z)}{h} r \sin \frac{2\pi n z}{h}$$

It looks something like this:

The length of the tinsel you'll need is just:

$$L = \frac{h}{2}\sqrt{1 + r^2\left(1 + (2\pi w)^2\right)} + \frac{h(1 + r^2)}{4\pi w r}\sinh^{-1}\left[\frac{2\pi w r}{\sqrt{1 + r^2}}\right]$$

We'll admit that this isn't the best-looking formula in the world. And, granted, applying this equation to calculate your tinsel length does require some knowledge of inverse hyperbolic functions. But that's still got to be easier than decorating your tree by eye, surely?

[2] This proof that there are only five platonic solids is based on the geometric argument in Euclid's *Elements*.

It goes something like this:

Every corner of a three-dimensional object must be the meeting point of at least three faces. Every platonic solid must have regular faces.

The adjacent faces at a corner must have a total internal angle of less than 360°.

Imagine constructing a corner from a flat piece of paper. You have to have some gap between two adjoining edges so that you can pull them together and make a convex three-dimensional shape.

The simplest regular face we can use has three edges, better known as the equilateral triangle. Let's begin by trying to construct some corners from those:

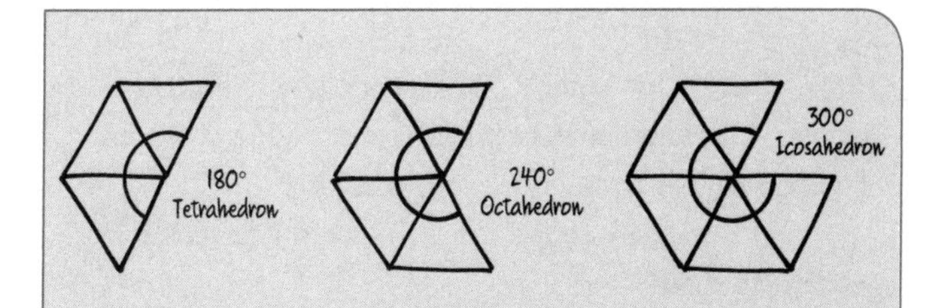

If three equilateral triangles meet at a corner (like in the image on the left) they have a total internal angle of 3 × 60° = 180°. This meets our criteria and, indeed, the corner illustrated is the one found in the tetrahedron.

If four regular triangles meet at a corner, as in the middle image, the total internal angle is 240°. Everything still works and this can be shaped to form the corners found in an octahedron.

Five regular triangles, like the image on the right, make the corners found on an icosahedron.

If you add one more triangle, however, the internal angle sums to 360° and – no matter what way you bend the paper – you can't make a corner. It will remain perfectly flat.

Moving on to four regular edges and hence square faces, only one platonic solid is possible:

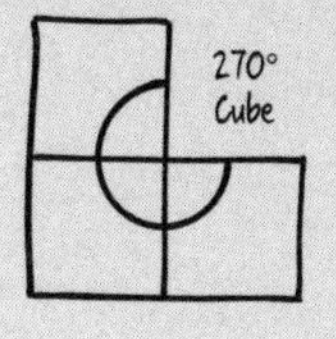

This is, of course, the corner you'll find in a cube. Add another square and you'll end up with an unbendable internal angle of 360° again.

A five-sided shape, too, has only one possible platonic solid:

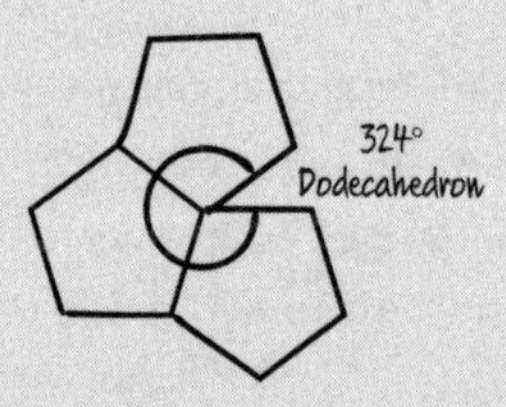

The internal angles of three pentagons will sum to 324° and the shape pictured will fold to give you the corners found on a dodecahedron.

Add another pentagon and you'll go past the magic number of 360° again.

And this is where our hunt for platonic solids ends.

Three six-sided regular polygons (hexagons) already add up to 360°. It doesn't matter how many more shapes you try, heptagons, octagons, decagons . . . three of them together will never dip below that crucial 360° angle again.

Hence there are, and only ever will be, exactly five platonic solids.

CHAPTER 3

Buying presents

What's the point of Christmas presents?

Every year, you can guarantee that at least half the gifts you receive will end up gathering dust in the back of a cupboard. Kitchen utensils designed to perform outlandishly specific tasks that have never been attempted outside a Michelin-starred restaurant. Bottles of perfume or aftershave from people who really have no business deciding what you should smell like. Novelty Christmas books you will flick through once and never open again . . .

We spend serious money on these things – about £600 per household in the UK,* making an eye-watering total of £16 billion a year† – but is it actually worth it? Why have we collectively decided to shower unwanted gifts on one another rather than, say, using the money to fund the NHS from Boxing Day through to 12 February?‡

* A. Farmer, 'British households' expected Christmas spending falls to £796', *YouGov*, 2015.

† There were 27 million households in the UK in 2015 (source: ONS).

‡ Based on an NHS budget of £120 billion per year (source: The King's Fund). For the record, the book you're holding is worth about 3 milliseconds of NHS funding.

Yeah, yeah. It's all about *showing people you care*. But that's such a frustratingly imprecise concept. Where's the analytical rigour? Surely maths can shed some light on how much we should really be shelling out on our loved ones.

To understand the mathematics of gift buying, we need to weigh the costs against the benefits. The cost of a gift is nice and easy to quantify, of course: just take a look at the price tag. The benefits are more tricky. People offer presents because it makes them feel good, and because it makes people they care about feel good too, but how do you put a value on that warm fuzzy feeling?

This problem – comparing outcomes that offer very different types of benefit – comes up all the time in a branch of maths called decision theory. Mathematicians judge the value of a benefit using something called *utility*, a sort of general feel-good factor, but measuring utility is not an exact science.

For instance, the utility you gain from offering a present might be represented by a graph like the one over the page.

The graph compares the utility you lose from the money you spend on a gift (in red) with the utility you gain from offering it (in grey).

The red line is easy to understand. The more money you have to fork out, the less you enjoy parting with it. The grey line is a bit more interesting though.

Since we're told that money can't buy happiness, we might imagine that the joy of offering a gift would

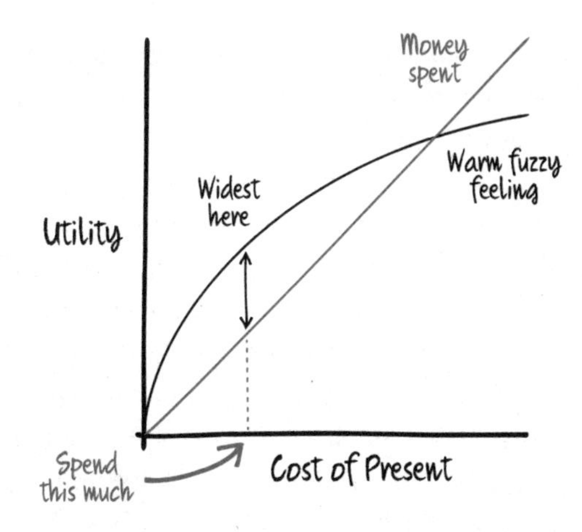

provide the same lump sum of utility no matter how much we spent on it. That wouldn't make much sense though, because if it were true we'd all offer presents worth next to nothing to get that hit of utility without any hit to our wallets.

The black line shows that the strength of your warm fuzzy feeling is actually related to the cost of the gift – the more valuable the present, the better you'll feel about offering it* – but the fact that the slope gets shallower as the cost increases shows that there are diminishing returns. At low prices, spending a bit more could make a big difference to your utility, but this extra benefit will drop off for more expensive gifts. If you're buying your dad a new watch, you might be able to get a much nicer one for £40 than for £20, but if you're torn between a gold Rolex for

* We're assuming you always get the best possible present for the money you spend.

£20,000 and a platinum one for £40,000, you've clearly gone mad and are unlikely to feel much better whichever one you go for.*

The vertical distance between the two lines tells you your net utility for buying a present at a given price. If the black line is above the red line, the warm fuzzy feeling is more valuable to you than the lost cash, so you make a net gain. However, if the red line is above the black line, you'll probably regret buying a present since the satisfaction of offering it isn't sufficient to outweigh your outlay. To maximize your utility, you should find the cost at which the black line is furthest above the red line and spend the corresponding amount.

We're not talking about precise measurements here. That's why there's no scale on the graph. It's only the relative size of costs and benefits that matters, and that will vary from person to person.

Take Great Aunt Hilda, for example. On the one hand, she is rich. If she wanted to, she could offer you all the presents listed in 'The Twelve Days of Christmas', and still have change for a family pack of frankincense. On the other hand, she is mean. This is a woman who

* There are a couple of other simplifications in the graph. First, we're assuming there's no benefit in offering a gift that you didn't have to pay for, which is not necessarily true, but doesn't affect our argument. Second, the way we have drawn the graph suggests that the relationship between money and utility is linear: double the money spent equals double the utility lost. The usual assumption is that money actually becomes less valuable to you the more of it you have (a pound is worth more to a struggling student than to a millionaire footballer), but so long as we scale everything else to match, this shouldn't make a difference either.

won't even share her cracker joke and always gives herself the piece of Christmas pudding with the coin in it.

Aunt Hilda's utility graph for offering presents would look something like this:

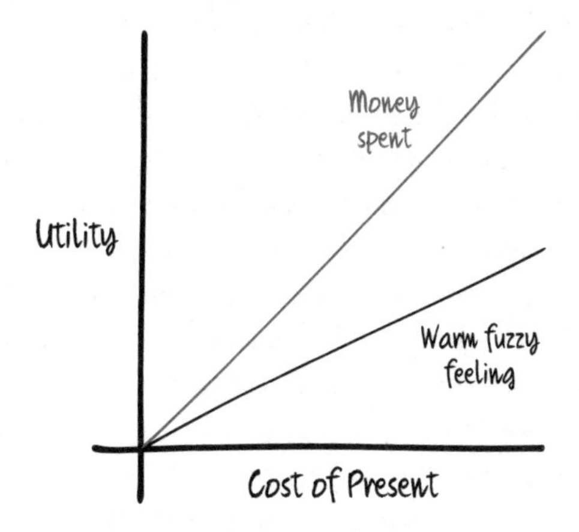

The old misery-guts values her money at double the value of offering a gift, so any pleasure she might get from buying you a present would be far outweighed by her disgust at having to open her purse. Naturally, you're equally unenthusiastic about getting anything for her, so we can suppose that when it comes to buying presents for Hilda, your graph looks much the same as hers.*

* To keep things simple, the benefit of buying a gift is now represented by a straight line. There are no diminishing returns for more expensive gifts. There's also nothing special about the fact that we've made Great Aunt Hilda value money twice as much as gift giving. It could have been three times or ten times as much and the analysis wouldn't change.

With the black line below the red line for all possible prices of present, there seems to be nothing for either of you to gain from offering a gift. Even if you're overcome by a sudden bout of Christmas cheer and do decide to get something for your least favourite auntie, there's no incentive for her to return the favour; quite the opposite, in fact. By accepting a present from you and offering nothing in return, she stands to make a tidy profit. On the face of it, then, it looks like you should just cross her off your list and be done with it.

Maybe things aren't so simple, though. Now that we're considering the effect of your actions and those of your aunt simultaneously, we've ventured out of the well-tilled field of decision theory and into the untamed wilds of game theory, the mathematics of interaction and competition.

Let's take another look at the possible results of an exchange of gifts with Hilda, then – but now we'll include the utility of any present you might receive from her (however unlikely it might seem that she will give you one).* We'll also assume that there is some maximum amount that you can afford to spend.

Your total utility (and your aunt's) is given by the following equation:

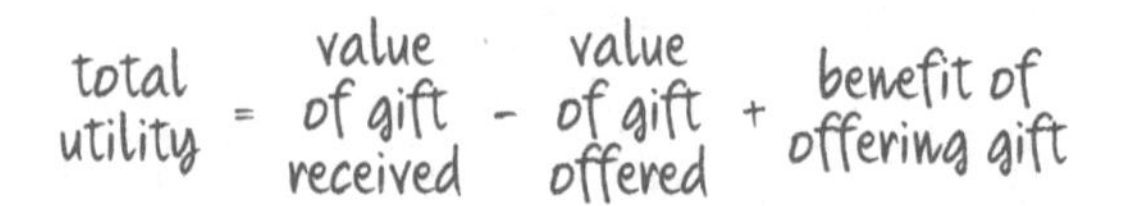

* We'll assume the utility of a gift received is equal to the utility of the money that was paid for it, since you could return it or sell it on if you didn't like it.

And since we're supposing that the benefit is equal to half the gift's value (in terms of utility), we can simplify this equation a little, to give us:

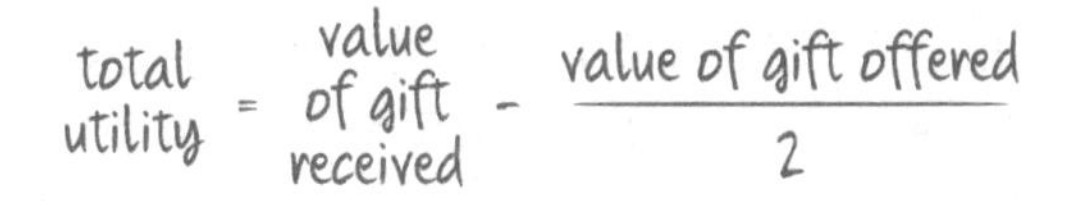

Using this equation, we can visualize all the possible outcomes in a single diagram:

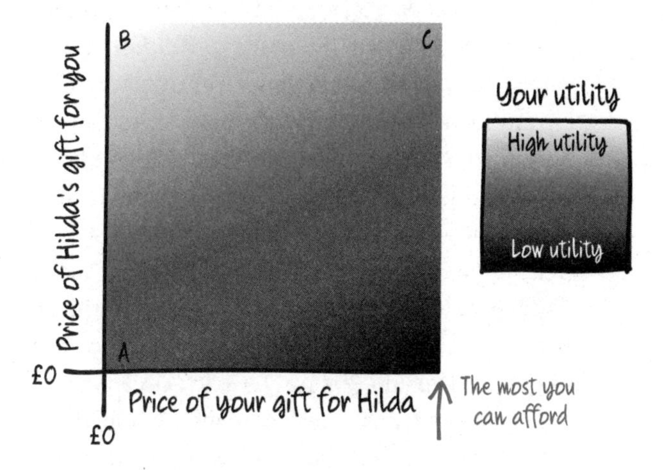

We've already seen that there's no incentive for either of you to spend any money on each other, so we're expecting Outcome A, down in the bottom left corner (dark red). From a selfish point of view, you'd prefer Outcome B in the top left corner (white), where you receive an expensive gift and offer nothing in return, but there's obviously no way you could persuade Hilda to go for that.

More interesting is Outcome C in the top right corner (pink). At that point, you offer as expensive a gift as you can and Hilda matches it. This is better than Outcome A for both you *and* your aunt, because while the value of your gifts cancels out, in Outcome C you also each get the benefit associated with offering your expensive presents.

The problem is how to achieve this outcome. Since you can't know how much your aunt has spent before you open your presents on Christmas morning – by which time you'll already have offered your gift – she's sure to try to maximize her utility by reneging on any deal you attempt to make. After all, there's nothing to stop her from gift-wrapping an empty box.

Luckily, there is one key fact that should allow you to extract a gift from the stingy old girl in spite of her treacherous nature. While it can't be Christmas every day, it is at least Christmas every *year*, with the big day putting in a dependably punctual appearance between Christmas Eve and Boxing Day each December.

This fact turns your whole interaction with Hilda on its head. When deciding whether to buy you a gift, Hilda can't focus purely on what will happen this year; she'll have to think about how you might behave in future years too, and this opens the door to a strategy that could ensure altogether merrier Christmases for both of you for many years to come.

In the first year, buy your aunt the most expensive gift that you can afford. Then, every year after that, buy her a gift of the same value as the present you received from her the previous year (up to your maximum spend). Also, crucially, explain this strategy to her in advance.

With this approach, it is in your aunt's best interests to match your spending. If she tries to exploit your 'generosity' by offering a cheaper gift than you, she will have to take a corresponding hit the following Christmas. Even if she doesn't believe you, it shouldn't take more than a year for her to realize you mean business, at which point her only sensible course of action will be to start offering expensive presents and reap the mutual rewards, thus guaranteeing that you'll end up with Outcome C every year.[1]

There is one catch, though. This strategy only works if you can be fairly sure you'll both be around to offer and receive presents when the next Christmas rolls along. Otherwise, you're back to a one-off interaction with no incentive to be generous. So, er . . . how to put this delicately . . . If you think that dear old Hilda is looking a little infirm, it may be better to try to exploit the situation by breaking your word and not buying her a gift after all, rather than counting on years of future gifting happiness that she may not be able to share with you.*

If you're lucky (and as cold and heartless as Frosty the Snowman), your aunt will have a more

* On the basis of the utility calculations in this chapter (where you and Hilda value offering gifts to each other at half the utility of their monetary value), you should avoid buying a present if you think your aunt's chances of seeing another Christmas are less than fifty–fifty.

favourable view of her life expectancy than you do, and so she'll go ahead and offer you an expensive gift as intended, resulting in the highly lucrative Outcome B. Then, if this Christmas does tragically prove to be her last, your profound sense of loss will be offset by the knowledge that you at least prised as much out of her as possible before she shuffled off.

Those are our top tips, then. Spend what you like on your nearest and dearest – your judgement is probably as good as any equation – but if you do happen to know anyone whose generosity is somewhat less than legendary, tell them you'll be matching the value of their previous year's gift and then spend as much on them as you and your credit card can stomach. However, if you have your doubts about whether they'll make it to next Christmas (or whether you'll still be in touch, at least), you're probably better off holding on to your money and hoping for the best.

And if that doesn't fill you with Christmas cheer, there's surely no hope for you.

ENDNOTE

[1] The Christmas present buying interaction between you and Hilda is a version of the Prisoner's Dilemma, a mathematical game that has been used to explain how cooperation can arise between essentially selfish individuals, from politicians competing for influence to animals competing for territory. The strategy of repeating your opponent's previous action is called 'Tit for Tat', and it is known to be one of the most effective ways to play, frequently outperforming much more complicated approaches in tournaments held between competing computer programs. You can find out more in Robert Axelrod's book *The Evolution of Cooperation* (1984; new edn 2006).

Although 'Tit for Tat' has mostly been studied in connection with the simpler *discrete* version of the game, in which players have only two options – to cooperate or to 'defect' – research suggests that it is also a good strategy for this *continuous* version of the game, in which participants can choose their precise level of cooperation (how much to spend). See J-B. André and T. Day, 'Perfect reciprocity is the only evolutionarily stable strategy in the continuous iterated prisoner's dilemma', *Journal of Theoretical Biology*, 2007, vol. 247, no. 1, pp. 11–22.

CHAPTER 4

Secret Santa

No matter how full of Christmas cheer you might be, if you work in an office there is one task associated with the season that everyone wants to avoid. No, we don't mean facing up to how much you shamed yourself at the Christmas party . . . We're talking about being lumbered with organizing the office Secret Santa.

If you're not already familiar with it, Secret Santa is a pretty cunning idea, dreamed up to save us all from wasting money on gifts for people we don't actually like, but feel obliged to buy for anyway. Participants put their names in a hat,* then each picks out a name at random, without revealing it to anyone else. Everyone then buys a present for the person they selected and the gifts are delivered without revealing who bought what.

Aside from only having to shell out for one gift, the beauty of the system is that you're basically free to offer any old tat while remaining gloriously anonymous. Novelty reindeer antlers, a wonky pumpkin mask (heavily discounted from Hallowe'en), a roll of parcel tape accompanied with the handy note 'Apply to mouth' . . . anything goes, really. There's no

* Make sure it's a Santa hat. Otherwise you're really not making the effort.

need for thought or tact, since any resulting resentment won't be directed at you but will be diffused across all participants, thus preventing irreparable damage to working relationships and allowing business as usual to resume when everyone manages to stagger back to work in the New Year.

As the organizer then, you'll be providing an important public service. You'd better make sure you get it right. It all seems pretty simple, though, doesn't it? Names. Hat. What could possibly go wrong?

Well . . .

From a mathematician's point of view, all that business of picking names is just a fancy way of creating a *permutation* of the participants. A rule that uniquely associates every buyer with a buyee.

Actually, you need a bit more than this, because a permutation can match a person with themselves. You don't want anyone to be their own Secret Santa, though. Then they might get themselves a gift they actually wanted, and that's not the point at all.

What you actually need is a *derangement*, a special kind of permutation where this self-matching doesn't happen. This is where the headaches really begin.

The usual way to stop people being matched with themselves is for anyone who does pick out their own name to return it to the hat immediately and pick again.

Seems like common sense, but there's a catch. What if the last person is left with their own name?

If this happens, there's no obvious solution. You can't just swap the name with someone else's, because then the last person could find out who was buying them a gift (and the person they swapped with would know who the last person was buying for too). Not-so-secret Santa.

In this case, the only real solution is to start the whole process all over again.

Luckily, with enough people, the probability of this happening is quite small – about 4% with 20 participants, and it gets smaller the more gift buyers you have* – but even if you're prepared to take that risk, this doesn't solve your problems. Not by a long way.

Because what if the second-to-last person picks their own name? They'll have to put it back and choose again, but then they'll know for sure that the last person is their Secret Santa.

And things aren't much better if the third-from-last person picks their own name. In that case, after they've returned their own name to the hat, the last two people will know the third-from-last person is still available to buy a gift for, so whichever of them does not pick that name will know that the other did. More generally, anyone who has to return their name to the hat has gained some information about their Secret Santa, since they know that it can't be anyone

* You might have thought that with n people, the probability of the last person picking themselves would be $1/n$, but it's actually less than that. Since the earlier pickers can't pick themselves, it's more likely than you might expect that the last person's name will already have been chosen. The actual formula for this probability is pretty horrendous, but if you really want to have a look at it, head to *The On-line Encyclopedia of Integer Sequences*: oeis.org/A136300.

who has already drawn. Even if the participants don't know the order in which they pick, the organizer does, so the secrecy is still compromised.

This 'common sense' system is suddenly looking pretty shaky.

There's another, more subtle problem with this setup too.

Your chance of being the Secret Santa for Rita on the front desk should be the same as the chance you'll end up buying for Barry in the post room. But using this system, that won't be true at all. For example, the last person to choose is always more likely to end up Secret Santa-ing for the second-to-last person than for the first.*

The tree diagram opposite shows how this works with just three people.

The numbers on the branches of the tree – either ½ or 1 in this case – are the probabilities that a particular name will be chosen at each stage. To see how the round of Secret Santa works, you move down the diagram.

Morris is the first to pick, and he has a fifty–fifty chance of picking either 'Doris' or 'Boris' (if he picks his own name, he will have to put it back and redraw, so we can ignore that possibility). If he picked 'Doris'

* In mathematical terms, we say that this name-replacement approach does not choose *uniformly* over all possible derangements. A full explanation of why this happens isn't easy. Loosely, though, you can see that the last person is more likely to pick out the name of the second-to-last person, because this will happen both in the case where they are the only one to pick this name, and also in the case where the second-to-last person picks their own name and returns it.

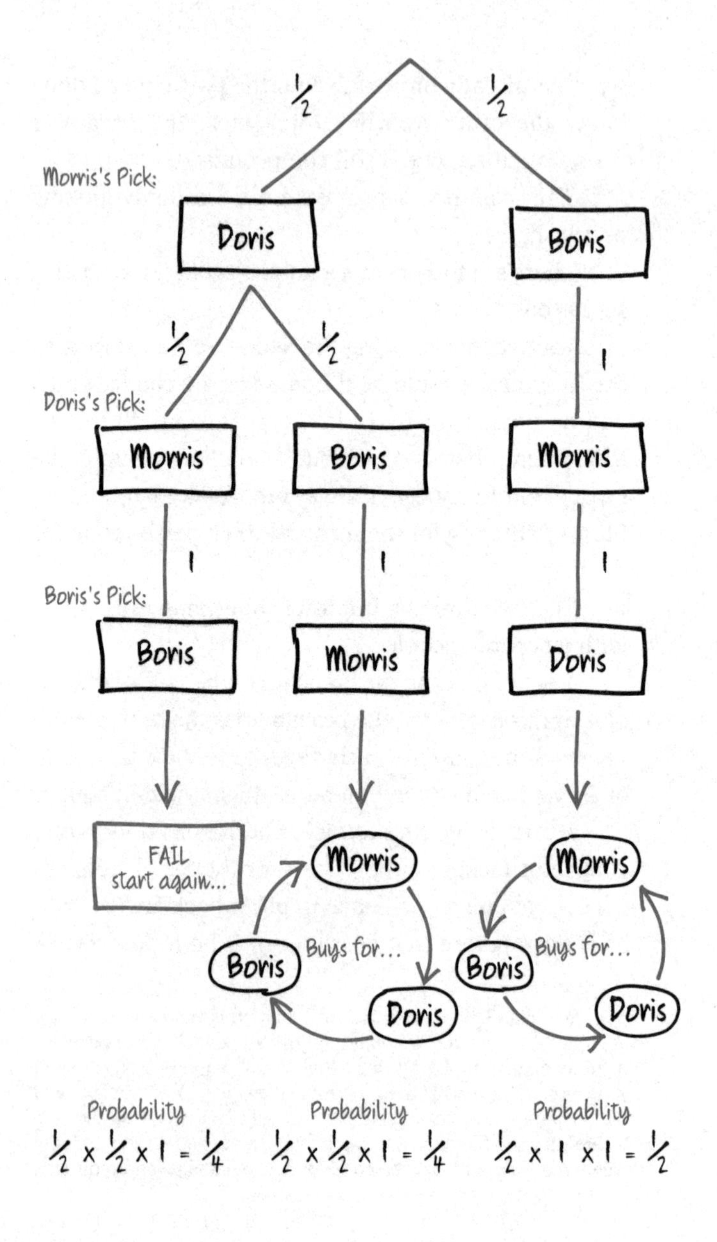
1/2
1/2
Morris's Pick:
Doris
Boris
1/2
1/2
1
Doris's Pick:
Morris
Boris
Morris
1
1
1
Boris's Pick:
Boris
Morris
Doris
FAIL
start again...
Morris
Boris
Buys for...
Doris
Morris
Boris
Buys for...
Doris
Probability
1/2 x 1/2 x 1 = 1/4
Probability
1/2 x 1/2 x 1 = 1/4
Probability
1/2 x 1 x 1 = 1/2

(the left-hand fork), then when her turn comes around, she also has two possible picks, 'Morris' or 'Boris', and again each is equally likely. However, if Morris had instead picked 'Boris' (the right-hand fork), then there is only one name left in the hat that Doris can pick out: 'Morris', so her probability of picking this name is 1. Either way, when it gets down to Boris's pick, there is only ever one name left.

To find the probability of each outcome, you multiply the probabilities on the branches you have followed. So we can see that the outcome in the bottom right, where Morris buys for Boris, who buys for Doris, who buys for Morris, has a probability of ½, while the outcome in the centre only has a probability of ¼, the same as the probability that Boris ends up picking his own name and you have to start all over again. This means that Boris is *twice as likely* to be Doris's Secret Santa as he is to be Morris's.

The standard Secret Santa system is fundamentally unfair. Even if you randomize the order in which people pick, you haven't solved the problem. Anyone who finds out the order (or even part of the order), will be able to work out which matches are more likely and the integrity of your Secret Santa will be seriously compromised.

To even things out, you could have everyone choose a name and only check at the end whether anyone had drawn their own, then start again from the beginning if they had. That would certainly be fairer. No one would know anyone else's Secret Santa, and every possible allocation would be equally likely – but there's a downside. That method comes with a whopping great failure rate of 37%.* Sure, you might get lucky and have things work out first time, but there's also a 5% chance you'll need at least four attempts,† by which point your colleagues may be less in the mood for buying presents and more inclined to strangle you with a piece of tinsel.

You're now running out of options. You could choose all the Secret Santas yourself, of course (or get a computer to do it for you), but picking the names is the only bit of Secret Santa that's remotely enjoyable. Besides, if the allocation is made behind closed doors, who knows what sort of conspiracy theories people might come up with. Santa-rigging accusations will fly, tears will be shed, and before long Christmas will be ruined for everyone.

To save Christmas, you need a system that's transparent (people pick their own match), fair (everyone has an equal chance of choosing everyone else), efficient (no need for redraws) and, above all, secret (no one has any information about who is matched or more likely to be matched, including you).

* Incredibly, this probability of failure – one minus the number of derangements over the number of permutations – barely changes, no matter how many participants there are. For groups of five or more, it's always just under 37%. In fact, this probability gets closer and closer to the value 1/e as the size of your group increases (e is a special number, like π, equal to about 2.718). You can see a video explaining this result on the YouTube channel of mathematician Presh Talwalkar: *MindYourDecisions*.

† The probability of three consecutive failures is $(1/e)^3$, which is just under 0.05.

And here is that method . . .

You'll need a set of cards like this one, one for each participant.

The number on the top and bottom parts of the card should be the same, but the numbers on each card should be different. Fold and unfold along the dotted line, so that the fold is visible from the back of each card.

The problem with the standard method is that you're creating a derangement as the names are drawn, a process that, as we've seen, is unfair and can go wrong. To get around this issue, we're going to use the cards to create our derangement in advance, then use the hat to place people into it at random.

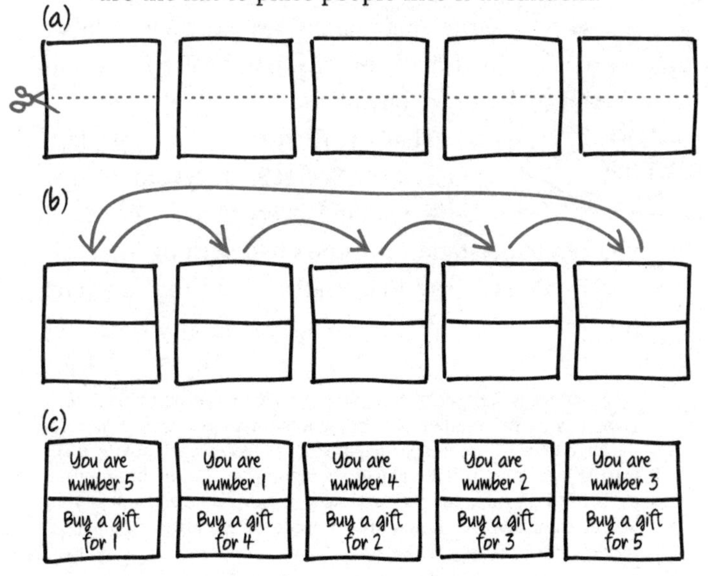

First, shuffle the cards and lay them in a row in front of you, face down. You should have no idea what order they are in. Now cut along the folds of each card, keeping all the pieces face down and in their places (a).

Next, move the top part of every card one place to the right and move the top of the last card back to the start (b). If you turned the cards over now, you would see that they describe a perfect derangement,* but with numbers instead of names (c). You don't need to turn them over, though; just stick or clip them back together and put them into your chosen sorting hat.

All you have to do now is to hand the hat round for people to pick out a card, and then stick a piece of paper on the nearest notice board where everyone can write their name next to their number and look up the name of the person they're buying for.

That's all there is to it. A totally fair, totally *secret* Secret Santa, guaranteed to work first time, every time . . . Provided you can understand everyone's handwriting, that is.

Don't ever say that maths never did anything for you . . .

* This process randomly generates a *cycle* of the numbers 1 to n (where n is the number of participants). This is a special kind of derangement where the participants form a loop, each buying a gift for the next one in line. This does mean that the method doesn't *quite* fulfil all our criteria, because you can never end up as your Secret Santa's Secret Santa, but we reckon that's probably a benefit rather than a drawback.

CHAPTER 5

Wrapping presents

Everyone loves a beautifully wrapped present. Even the most slapdash and ill-considered of gifts can be made to look attractive if it's hidden away behind a layer of neatly folded and cheerfully illustrated paper. That shiny wrapping, that dainty gift tag and that ridiculous big stick-on bow. It's all just one big guilty apology. 'Sorry I couldn't be bothered to take the time to come up with anything better than this generic tat; please accept my two minutes of folding and sticking as a worthy token of my shame.'

Except, of course, many of us can't manage to get the folding and sticking part right either. For the first present you cut too much paper, for the second you cut too little, and by the time you reach the end you've run out of paper altogether and have to try to patch together whatever ragtag bunch of offcuts you have left. The resulting parade of sad and crumpled packages, tangled in twisted sellotape, would bring a tear to the eye of even the most indifferent of Christmas killjoys.

What most people don't appreciate is that wrapping is a game of two halves. The efficient swaddling of your Christmas offerings doesn't start when you pick up those scissors at six o'clock on

Christmas Eve. It actually begins a couple of hours earlier, in the shops. Mathematically speaking, the first secret of good present wrapping is to make sure you only buy presents with shapes that can be efficiently wrapped.

Wrapping paper doesn't come cheap, so it stands to reason you'll want to choose gifts that require as little of it as possible. Naturally, the smaller the present, the less paper required; but if we suppose that there is some minimum acceptable volume of gift you can offer without looking like a total Scrooge, what shape of present should you seek out to minimize your outlay on wrapping paper?*

The first key concept here is the *surface area to volume ratio* of your gift: the surface area divided by the volume. For our purposes, it will be simplest to think of this number as the surface area (in square metres, say) of a solid shape that has the same shape as your present, but with a volume of one unit (e.g. 1 metre cubed†). For example, each side of a cube of volume 1 m^3 must be 1 m long, and a cube has six faces, so its surface area is equal to $6 \times 1 \text{ m} \times 1 \text{ m} = 6 \text{ m}^2$ and its surface area to volume ratio is 6.[1]

* If you're thinking that the price of the gift itself is likely to be more significant than the cost of the paper in determining your total spend, you've obviously never received a present from either of us.

† It's probably fair to say that if you're in the habit of buying gifts with a volume of 1 cubic metre, you are unlikely to be the sort of person who needs to worry about the cost of their wrapping paper. In any case, the key point is that the base unit for the measurement of the surface area (in square units) and the volume (in cubic units) must be the same.

Miserly wrappers that we are, we want this ratio to be as small as possible, minimizing the surface area that must be covered with expensive paper, and for cuboidal gifts – that is, packages with six rectangular faces, not necessarily equal in area – it turns out that the cube's score of 6 can't be bettered.

Here's a chart to show how different cuboids shape up:

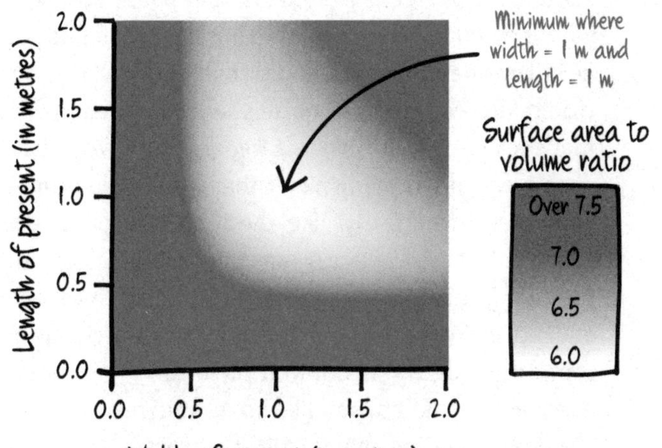

The surface area to volume ratio of a cuboidal present depends on its length and width,* so those are the measurements on the two axes. The depth of colour represents the value of the ratio for each pair of measurements, with the lowest value in white and the higher values in darker red.

* Since we're only comparing solid shapes with a volume of 1 m^3, once you know the length and width, the height is fixed.

From the graph, we can see that the ratio is smallest when the length and width of the box are both equal to 1 m. Since the volume is fixed at 1 m^3, the height must be 1 m too, confirming that a cube is our best bet.*

But hold on. Before you go crazy and order a bumper pack of Rubik's Cubes for your nearest and dearest, let's take a moment to consider. Thinking about surface area is a good starting point, but when you sit down to do the actual wrapping, you'll find you need considerably more paper than your calculations would suggest, because of all the useless flappy bits that stick out when you start folding. Sure, you could cut a piece of wrapping paper that would fold around your present perfectly, like the template for a cube given in Chapter 2, but if our objective is to save on paper, this is a pretty stupid idea, since you'll probably end up throwing away a lot of what you cut away round the edges. Plus, if you have time to measure out complicated shapes like that, you really should consider spending it with the people you're buying the presents for, rather than locking yourself away in your wrapping shed for most of December.

Let's be sensible, then, and limit ourselves to rectangular pieces of wrapping paper. The standard way to wrap a cuboidal present is to place it with

* Looking at a diagram isn't a proof, of course. If the best measurements were 1.01 m by 0.99 m, for instance, we couldn't tell the difference by eye. We *could* prove that the cube is best using a delightful bit of partial differential calculus, but we're guessing you'd rather we didn't. No calculus at Christmas, that's the rule, right?

its edges parallel to the edges of the paper and fold all the way round in one direction so that opposite edges of the paper meet in a straight line on the top, before folding in to cover the two ends that remain uncovered (see Diagram 1 below).

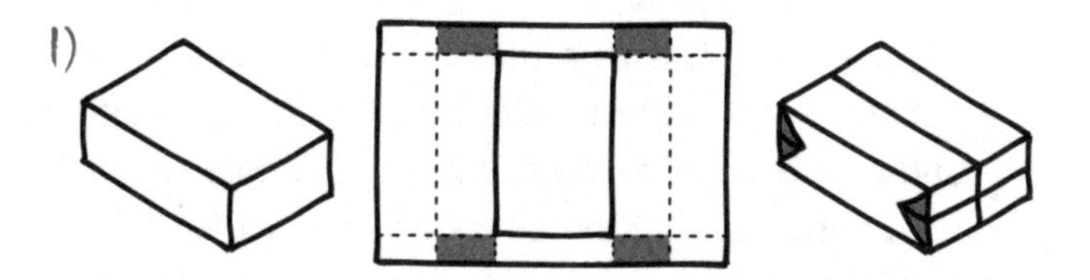

For maximum mathematical efficiency, you should measure out your paper so that one edge is equal to the sum of the length and the height of your present, while the other edge is equal to double the sum of the width and the height* (though, of course, you'll need to make the rectangle a tiny bit larger, to give yourself a bit of leeway). That way, everything should meet up perfectly, as in Diagram 1.

We've marked the excess paper in red on the diagrams. It's actually quite easy to see how much has been wasted, because you can imagine those four red rectangles being stuck together into two squares, each with side length equal to the height of your present. The wastage with this method is therefore equal to twice the square of the height.

There is also a diagonal method for wrapping cuboidal presents, one that smarmy wrappers claim is

* For maximum efficiency, the height must be the shortest of the three measurements. The width and length are interchangeable.

superior. This alternative technique is often presented as some sort of mystical secret, accessible only to the pure of heart, so powerful that those who master it are able to completely cover a fridge–freezer using only a postage stamp.

Sadly, diagonal wrapping is not a miraculous way to hack your Christmas. Mathematically speaking, it's actually a bit rubbish. While the technique may be sufficiently novel to dazzle the easily impressed, it is in fact less efficient than the standard method.

The diagonal approach uses a square piece of wrapping paper, with a side length equal to three-quarters of the sum of the width, the length and *double* the height of your gift. This produces a piece of paper that's a touch larger than necessary, so there's no need to factor in any additional overlap. Then you place your present in the centre, at a 45° angle, and fold the four corners over on to the top, as here in Diagram 2.*

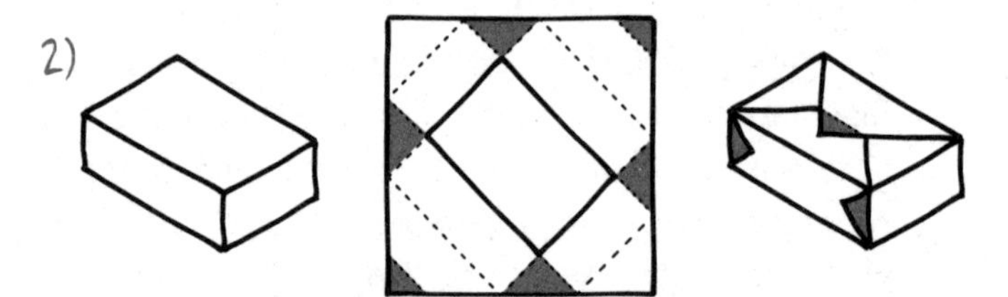

* The precise expression for the side length is $\sqrt{2}(h + w/2 + l/2)$ and our wastage calculations are based on this exact formula, rather than the approximation. The expression was first outlined by Dr Sara Santos ('Formula for the perfect present wrap', *Daily Mail*, 13 December 2005), who makes the point that, even if it doesn't actually save paper, the technique does require much less tape than the parallel approach. You can see a video of Dr Santos presenting the method on *The One Show* in 2012 on her YouTube channel: *pestantan*.

Once again, there will be four flaps of excess paper at the corners of the gift – triangles this time, marked in red again – and once again you could imagine these being stuck together to form two squares with side length equal to the present's height, creating the same amount of wastage as before. However, this time there are two more triangles of excess paper that overlap on the top of the package. Admittedly, if your present has a square base, this extra wastage disappears and the corners all meet in a tidy point (Diagram 3), but in terms of paper efficiency you're still no better off than with the basic method.

In fact, neither of these methods is truly the most efficient. Using a sufficiently long and thin strip of paper, you could wrap a cuboidal present with essentially no wastage at all by winding the paper round and round like a bandage.* However, unless your gift of choice is a 5,000-year-old Egyptian pharaoh, this is unlikely to be a very sensible option.

* E. Demaine et al., 'Folding flat silhouettes and wrapping polyhedral packages: new results in computational origami', *Computational Geometry*, 2000, vol. 16, no. 1, pp. 3–21.

Whether you go with the parallel or diagonal technique, one consequence of those flaps of excess paper is that, despite its low surface area to volume ratio, a cube is *not* the most efficiently wrappable cuboid. While a flatter box has proportionally more surface area, it also wastes less paper at the corners. For the most economical wrapping, you should spend your Christmas shopping trip hunting down gifts in square boxes, whose height is half the side length of their base, like the one in Diagram 3. A box like that needs almost 11% less wrapping paper than a cube of the same volume.

The technique shown in Diagram 1 can also be adapted for other solid shapes (though it's not always the most efficient). To wrap a cylinder, you should cut a piece of wrapping paper whose length is equal to the sum of the cylinder's length and diameter, and whose width is equal to the circumference, as in Diagram 4.

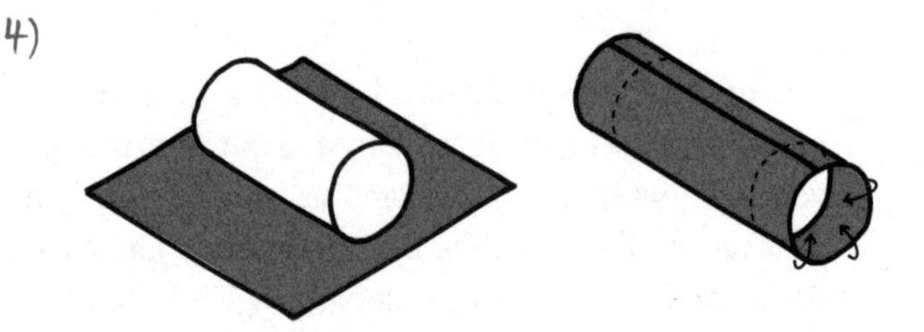

For an equilateral triangular prism (giant Toblerone, anyone?), you need a piece whose length is equal to the length of the prism plus two-thirds of the vertical height and whose width is equal to the

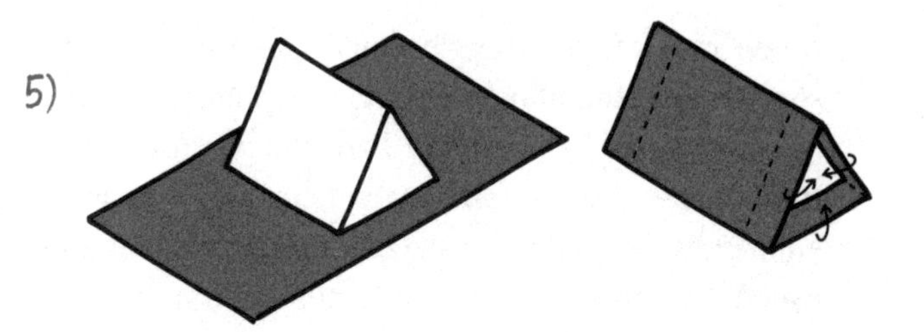

perimeter of the triangular face. Diagram 5 shows how it works.[*2]

In terms of surface area to volume ratio, though, there is one shape that leaves all others in the shade. A sphere has barely 80% as much wrappable surface as a cube of the same volume, making it the most efficient solid shape of all. Logically then, spherical presents should be the holy grail of Christmas cheapskates everywhere, offering an unparalleled opportunity for wrapping frugality.

Except . . .

While a sphere might have the least surface area for a given volume, if you've ever tried to gift-wrap a football you'll have discovered that actually covering that theoretically efficient surface with a practically efficient amount of paper is easier said than done. No sooner do you get the paper smooth in one place than it crinkles up somewhere else, and after

* This method gives you just enough paper to cover each end of the prism. For a slightly different way to wrap triangular prisms, along with demonstrations of wrapping techniques for several other shapes, you could do worse than have a look at 'Mathematical present wrapping', a video from mathematician Katie Steckles, available via her YouTube channel: *st3cks*.

a few minutes of struggle you'll likely end up with something resembling a scrunched-up mess.

You shouldn't feel too bad about your failure, though, because wrapping a sphere neatly is mathematically impossible. It all comes down to something called the *Gaussian curvature*,* a number that indicates how curvy the sphere (or any other shape) is at a particular point on its surface. The Gaussian curvature is actually calculated by multiplying two numbers together. If you think of the sphere as the Earth and you imagine you're standing on the equator, then one of these two numbers would measure the curviness in the north–south direction, the other in the east–west direction.

A sphere curves away from you everywhere, in all directions, so the Gaussian curvature is positive all over the surface. A piece of wrapping paper, on the other hand, is (initially) flat everywhere, so the curviness in both directions is zero and the Gaussian curvature is zero everywhere. A cylinder has zero Gaussian curvature too, since while it is curved in one direction (around) it is flat in the other (along), and a positive number times zero is still zero.

The killer fact – look away now if you count perfect sphere wrapping as one of your life goals – is that the Gaussian curvature of a surface will always

* A. Bhatia, 'How a 19th century math genius taught us the best way to hold a pizza slice', *Wired*, 5 September 2014.

stay the same when you bend or fold it.* No matter how much you contort your wrapping paper, if you make it curvy in one direction, it has to stay flat in the other, so that when the two measures are multiplied together, one of them will be zero and thus the result zero too. Whatever you do to it, the paper will stubbornly maintain the zero Gaussian curvature it started off with.

Tragically, this means that while you should have no trouble wrapping a festive tin of beans, your paper will never be able to match the three-dimensional curviness of a tangerine.†

All is not lost, though, because if you can put up with a bit of crinkle, there are two different fudges that do allow you to wrap a sphere in a rectangular piece of paper in an *almost* tidy fashion. Although the measurements needed for the two methods are different, the amount of paper used is the same, so which one you choose is really just a question of aesthetics.‡

The first method (Diagram 6) requires a square

* You can change the Gaussian curvature of a surface if you stretch it, but wrapping paper is not very stretchy. Wrapping your gift in latex is an option, though . . . We'll leave you to decide whether to follow up on that.

† You can't wrap a horse's saddle either, though for a slightly different reason. Because a saddle curves upwards in one direction (along the horse) and down in the other (around the horse), its Gaussian curvature is calculated by multiplying a positive curvature by a negative curvature, giving a negative result. Again, there is no way to reproduce this negative Gaussian curvature with a flat piece of wrapping paper. Looks like Blossom will have to make do with an unwrapped gift this year, eh?

‡ O. Aichholzer et al., 'Wrapping spheres with flat paper', *Computational Geometry*, 2009, vol. 42, no. 8, pp. 748–57.

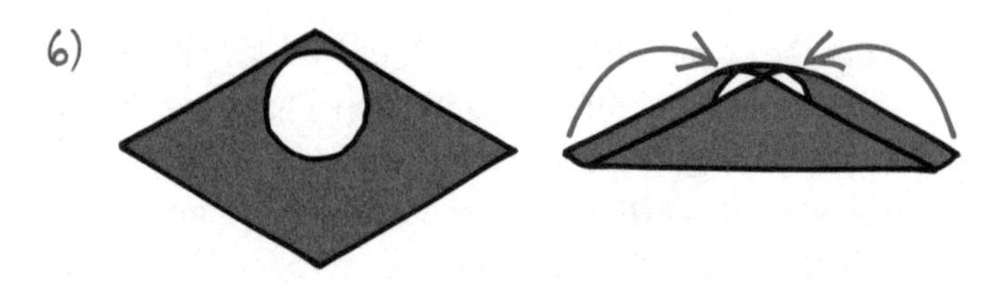

piece of paper with side length of two and a quarter times the diameter.* If you put your gift in the middle of this piece of paper, you should just be able to bring the four corners over to meet on the top. Then you can simply press down the unavoidable big sticky-out bits and hope for the best.

Alternatively (see Diagram 7 below), you can cut a rectangle of paper, with one side equal to the circumference of your present and the other side equal to half that length.† Using this method, you roll the gift up in a cylinder of paper, then gently press in the ends. You'll have to make sure it's right in the middle though, because if you've measured accurately there should only be just enough paper to cover it.

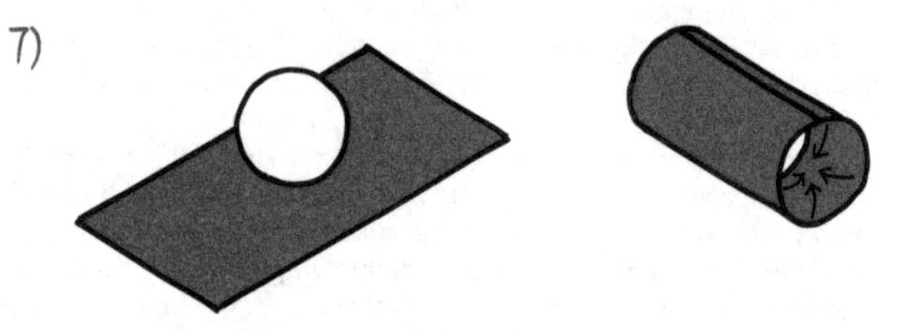

* The precise side length is $\pi d/\sqrt{2}$, roughly 2.22 times the diameter, d, so our suggestion gives you a bit of margin for overlap.
† In terms of the diameter d, the dimensions are πd and $\pi d/2$.

Although these methods are the best you can do with a rectangular piece of paper, they still fall well short of wrapping perfection. Despite the sphere having almost 20% less surface area than a cube of the same volume, our wrapping techniques use only 5% less paper and the end results are not exactly works of art either. We must be able to do better.

Well, as it happens, you *can*, but you'll need a lot of patience. There is a family of shapes called *petal wrappings* that can wrap a sphere as beautifully and as efficiently as you like (Diagram 8).* The only catch is that if you want to achieve maximum efficiency you'll need to cut a piece of wrapping paper with an infinite number of petals. So while you are welcome to have a go at this if you like . . .

. . . it might be easier to put your ball in a box.

8)

* Aichholzer et al., 'Wrapping spheres with flat paper'.

ENDNOTES

[1] Surface area to volume ratio is actually a tricky sort of quantity, because it doesn't remain constant as the size of a solid shape increases. For instance, although a cube with side length 1 metre has a surface area to volume ratio of 6, if you double the length of each side to 2 metres, the new cube will have a volume of 8 cubic metres and a surface area of 24 square metres, giving a ratio of just 3 (24 divided by 8). The surface area to volume ratio is not fixed for each shape. Thankfully, since we are only comparing presents of the same volume, we don't need to worry about this. Oh, and strictly speaking, the units for the ratio should be measured in inverse length, since it's a square unit divided by a cubic one, but hey; Christmas is too short to worry about that sort of thing.

[2] Diagrams 4 and 5 actually show the most efficiently wrappable cylinder and triangular prism using the method depicted. Here are some extra mathematical notes on wrapping cylinders and prisms with this technique, for completists:

- The wastage of the wrapping methods shown in Diagrams 4 and 5 is always equal to twice the area of the end of the triangular prism or cylinder.
- For a cylinder, the lowest possible surface area to volume ratio is achieved when the length is

equal to the diameter. But the most efficient cylinder to wrap, requiring over 5% less paper, is actually twice the length of its diameter.

- The triangular prism with the lowest surface area to volume ratio has an equilateral cross-section and is $1/\sqrt{3}$ times as long as the side length of the triangle. The most efficiently wrappable prism is also equilateral, but is $2/\sqrt{3}$ times as long as the side length. Again, it requires over 5% less paper than the prism with the minimal ratio.
- You can prove that an equilateral triangular prism is the most efficiently wrappable by showing that the amount of paper required to wrap any triangular prism always increases as you increase the perimeter of the triangle (keeping the volume and cross-sectional area constant). Since an equilateral triangle has the shortest possible perimeter for a given area, it must be the most efficient.
- If the triangular prism is not equilateral, calculating the size of the piece of wrapping paper needed for the method shown in Diagram 5 is a bit more complicated. The width is still equal to the perimeter of the triangle, but the length should be equal to the length of the prism, plus four times the area of the triangle divided by its perimeter. The bit of paper sticking out at each end of the prism will then be just long enough to reach the *incentre* of the triangular face, the point furthest from any of the edges.

CHAPTER 6

Cooking turkey

So you've decided to host your family for Christmas this year. On the plus side, this means you don't have to stay at Grandma's (always freezing, smells of cats) or trek over to your sister's (where God forbid you should spill any red wine on the sofa). On the downside, it also means you're now in charge of cooking the Christmas dinner.

Even though you're a bit of a dab hand at meat-based finger food, a platter full of pigs in blankets probably isn't going to cut it. If you want your Christmas lunch to go down as a wild success, you're going to have to master the centrepiece of the festive feast and tackle the turkey.

Cooking a turkey poses quite a culinary challenge. If you look past the feathers, wings and beak for a moment,* you'll find that it's made up of 60% water, 20% protein (muscle fibres, connective tissues) and 20% fats. Heating the muscle fibres causes the proteins to coil up and the meat to contract. Initially this makes the meat nice and tender, but heat the fibres too much and they'll start to clump together

* Ignoring parts of the turkey will become a recurring theme of this chapter. Just you watch . . .

and *coagulate*, making the meat tough. Worse, the more the meat is cooked, the more of the water that's trapped in the proteins will escape, making it taste dry and horrible.

On the flip side, the proteins in the connective tissue, which hooks the muscles up to the bones, unwind or *denature* from tough collagen into soft gelatin when heated, making the meat edible.

The secret to a good roast turkey, then, is to heat it enough to denature the collagen but not enough to coagulate the muscle fibres too much.*

If that isn't tricky enough, this balance is going to change in different parts of the bird. Generally, breast is best when cooked at high temperatures for a short amount of time, while the leg is tastiest when slow cooked at a lower temperature. But unless you want to ruin everyone's fun and hack the bird up to cook it in bits, the turkey needs to remain intact. Each part will be left cooking in the oven for the same amount of time.

Still sure you wanted to host Christmas this year?

Yes? Then fear not! Maths is here to help. Well. Jamie Oliver's maths anyway.

Jamie has distilled all that complexity down into a simple algorithm† to give you the time, *T*, needed to cook the perfect turkey:

* For more on the chemistry of the turkey see Roger Highfield's book *Can Reindeer Fly?* (1998).
† Source: jamieoliver.com.

(1) Cook for $T = 2W/3$ hours at 180°C where W is the weight of the bird in kilos.
(2) Cover the bird in foil until the final stages of cooking.
(3) Shove half a clementine up its bum.

Step 2 of this algorithm (OK, OK, 'recipe') stops too much water from escaping from the bird during cooking. Step 3 probably does the same thing. Although we like to think it's in the recipe just because it's a fun thing to do.

But Step 1 troubles us. We reckon we can do better than that. And seeing as how this recipe, humbly named 'The best turkey in the world', only managed 3 stars out of 5 on its online reviews when we last checked, we feel it's only right that we offer our mathematical services, free of charge,* to help Jamie's turkey live up to its title.

The objective here is to cook the turkey for the minimum amount of time required to get the centre of the bird up to the temperature at which it's safe to eat† – 75°C – but no longer, or its muscle fibres will start to spoil. To translate this into a mathematical problem, we first need to make a few minor quasi-realistic and totally justifiable assumptions.

If you squint at your plump Christmas turkey from an angle you *could* say that it is roughly spherical. True,

* Well. For the inexpensive price of £9.99. Or less if he waits until April and finds a copy of the book in a bargain bin at Poundland.
† Source: cooksrecipes.com's meat cooking temperature chart.

you may have to close your eyes almost completely and get down to an unnaturally oblique angle, but let's say we're talking about a particularly overfed and genetically modified ball in your roasting dish.

One more thing – no big deal, but we're also going to have to assume that there is no cavity in the middle of the turkey – also no bones, no giblets, no leg meat, no breast meat. Just a clean and simple faultlessly homogenous fleshy sphere. In short, a festive dish without any of the characteristics that might have once provided the turkey with any semblance of life. (Think Bernard Matthews turkey balls.)

This might seem a little extreme, but just go with us for a moment.

Ultimately the turkey heats up by conduction. Energy is transferred between neighbouring particles and diffuses down from areas of high temperature to the cooler regions until eventually everything reaches equilibrium and the turkey is the same temperature as the oven throughout (although you'll take it out long before that happens).

If your turkey was perfectly spherically symmetrical, the only thing that would matter for the cooking time would be how far away each bit of meat was from the centre. The turkey would essentially be no more than an infinite collection of lines of meat, all joined in the centre and pointing out radially in all directions, as far as the surface of the bird. The only way for heat to conduct through to the centre would be via that surface, which, if the turkey has radius R, would have an area of $4\pi R^2$. The bigger the bird, the bigger the surface is, but also the further away the cool centre is from the heat in the oven.

In this case, Fourier's law of conduction says that the cooking time *T* of the turkey will be equal to some constant *a* (which depends on how good the meat is at conducting heat) times the radius squared:

$$T = aR^2$$

But for a spherical turkey, the weight also has a simple* relationship to the radius:

$$W = 4\pi \rho R^3/3$$

where ρ is the density of the bird. With a bit of messing around, you can combine these two facts to get an alternative equation to Jamie's that, with an oven still set to 180°C, relates cooking time in hours, *T*, to the weight of the turkey in kilos, *W*:

$$T = 1.13W^{2/3}$$

The 1.13 that has cropped up in this version depends on the constant *a* and the density of your average bird and was determined by Stanford University's Pief Panofsky.† He found the value 'empirically', by which we think he means he ate a shedload of turkey and decided which tasted the nicest.

* Simplicity is in the eye of the beholder.
† For Panofsky's original equation, see symmetrymagazine.org. It originally related pounds to hours, but we have converted it to metric here.

It might not look that nice, but actually this equation does make good intuitive sense. Certainly a lot more than Jamie's. If a bird doubles in weight, it's going to get bigger in all three dimensions. But the only way the heat can get in is through the surface, which is only two-dimensional: hence the power of 2/3.

Below is a comparison between the two formulas. If we trust Panofsky, then Jamie's recipe will undercook smaller turkeys, overcook larger birds and only provide the tiniest window of perfection at about 4.9 kg.

Given the nature of our assumptions (globular mutant turkeys with no internal organs), some of you may be suspicious. But let's see if we can persuade you that there is some metaphorical meat on our mathematical bones.

Irrespective of Panofsky's formula, these assumptions allow us to write down a rather delicious-

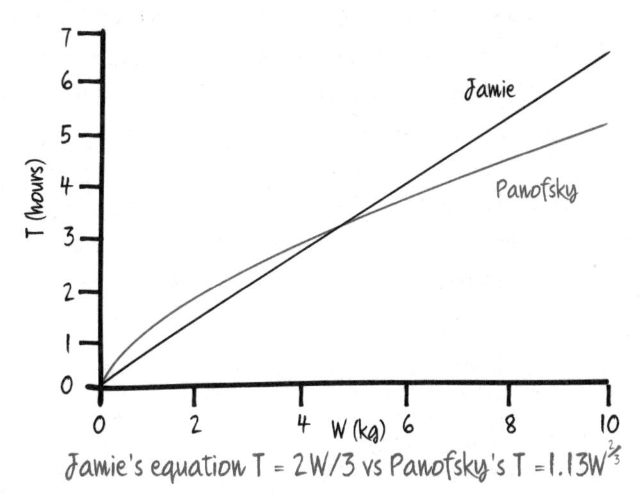

Jamie's equation $T = 2W/3$ vs Panofsky's $T = 1.13W^{\frac{2}{3}}$

looking equation – not just for the total time it will take for the bird to cook, but for the temperature ϕ inside the turkey at any point r at any time t.

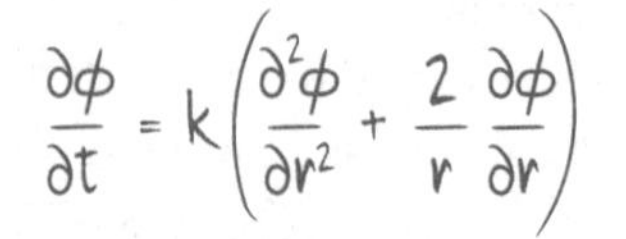

$$\frac{\partial\phi}{\partial t} = k\left(\frac{\partial^2\phi}{\partial r^2} + \frac{2}{r}\frac{\partial\phi}{\partial r}\right)$$

This is the heat equation in radial coordinates. The left-hand side dictates how the temperature changes over time. The right-hand side explains how the heat diffuses through the meat depending on how close you are to the centre of the turkey.[*][1]

The heat equation might look a little more intimidating than Jamie's to your average home cook, but unlike the Naked Chef's formula, this version can be easily tested in a culinary environment.

Empirically finding a direct relationship between weight and cooking time is tricky because you have to repeat the cooking process for a huge number of birds – all of which will have different dimensions and physical characteristics.

But to validate the heat equation,[†] you simply take a single turkey and monitor the temperature while it cooks. And that's exactly what we did.

So here's one we cooked earlier.

* The k here is the *thermal conductivity* of the turkey, while the ∂ symbol represents a *partial derivative*.

† We also require the additional (but less dramatic) assumptions that the turkey begins at the same temperature throughout and that the oven stays at a constant temperature while cooking.

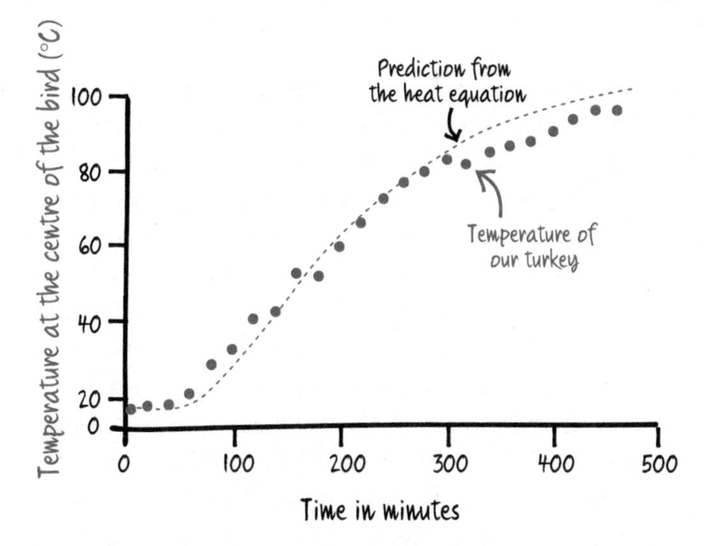

The circles on the graph are our temperatures from the centre of our 5kg bird,* cooked from room temperature (16°C), while the oven was set at 180°C. The dotted line is how the heat equation predicts the temperature will change over time.

In both prediction and experiment, it takes some time before any heat from the oven penetrates through to the middle. The bird then heats up steadily until it hits around 80°C; from then on the temperature continues to rise,† but now every extra 5°C takes progressively longer. (That's why when you're roasting dishes, the

* We set our oven to 180°C to match Jamie's recipe, but that equated to a temperature of only 120°C inside. Probably because we kept opening the oven every ten minutes.

† Although our bird was probably ready to eat after about three and a half hours, because we wanted to test the heat equation we continued to cook it past 100°C. In the end it tasted absolutely terrible. We hope you appreciate the sacrifices we made for this book. And the sacrifice the turkey made too, come to that.

closer you get to their being cooked the longer it seems to take to finally get there, and why Christmas dinner is always at least an hour later than advertised.)

The fit between the experiment and the prediction isn't perfect, but as you can see, they are in close agreement. The assumptions it took to get there might well have seemed like massive oversimplifications, and yet they have still allowed us to capture the essence of what's happening inside a real bird as it cooks.

Despite missing some minor nuanced details like 'bones', we have ended up with a very close approximation of reality in the form of a solvable mathematical equation. Which gives us a massive advantage when trying to plan our Christmas dinner.

But if you're still not happy, you can take a look at the work of Chang, Carpenter and Toledo,* who implemented a more advanced version of our heat equation, taking into account the thermal conductivity of the thigh, breast and cavities while modelling the effects of convection, radiation and surface energy loss, all leading up to the incredible discovery that:

> *Increasing oven temperature reduced baking time but resulted in breast temperature reaching the designated end-point much earlier than what would be required for thigh-joint temperature to reach the end-point.*

* H. Chang et al., 'Modeling Heat Transfer During Oven Roasting of Unstuffed Turkeys', *Journal of Food Science*, 1998, vol. 63, no. 2, pp. 257–61.

Alternatively, Travis Mikjaniec* used a three-dimensional computational fluid dynamics simulation of a cooking turkey to model the typical airspeed inside a normal convection oven and make a contour plot of the velocity of the air within a turkey's cavity to reach the immensely valuable and breathtakingly profound conclusion that adding stuffing does not impede the airflow.

(To be fair, the methods used by both these groups have many applications in other realms of science, from racing-car engineering to understanding the behaviour of *Salmonella*. And we've written a book about Santa. The views from this glass house really are spectacular.)

But if you're still not sure when it comes to your own Christmas dinner, might we suggest that you just use a meat thermometer. Or ask your grandma to do the cooking.

DISCLAIMER

Make sure all the juices are running clear and that your turkey is fully cooked before eating it. Remember – we're mathematicians. We live on Pot Noodles and instant coffee. We've just spent a chapter pretending that turkeys are spheres of uniform consistency. We really don't know what we're talking about, so please don't poison yourselves, OK?

* T. Mikjaniec, 'How to cook turkey like an engineer', deskeng.com, 2013.

ENDNOTE

[1] This concept of diffusion from one area to another isn't unique to turkey cooking; it's far more powerful than that. The heat equation provides the simplest description of this process and you'll find it popping up in some form or other across all the sciences, describing how fluids move and how the financial markets behave. Far from shoehorning one equation into lots of different scenarios, this is actually a great example of how thinking mathematically often helps to draw parallels in the underlying processes between seemingly unconnected areas.

Solving the heat equation in the particular case of our spherical turkey of radius *R* with oven temperature $\emptyset s$ and initial meat temperature $\emptyset o$ gives an expression for the temperature at the centre of the turkey $\emptyset c$ at time *t*:

$$\frac{\emptyset_c - \emptyset_s}{\emptyset_0 - \emptyset_s} = 2 \sum_{n=1}^{\infty} (-1)^{n+1} \exp\left(\frac{-k n^2 \pi^2 t}{R^2}\right)$$

Notice how the initial temperature of the bird affects the temperature at the centre at a particular time. If at all possible, you should try and get your turkey up to room temperature before cooking to reduce the risk of overcooking the breast.

EXTRA READING

J. Unsworth and F. Duarte, 'Heat diffusion in a solid sphere and Fourier theory: An elementary practical example', *American Journal of Physics*, 1979, vol. 47, no. 11, pp. 981–83.

CHAPTER 7

Cutting the cake

With the best will in the world, one turkey – no matter how mathematically perfect – does not really constitute a Christmas feast. So, assuming that you have managed to avoid poisoning your relatives (intentionally or otherwise), you will probably need to follow up your festive fowl with something sweet, and what better than a good old-fashioned Christmas cake?

You'd better tread carefully, though. In the volatile environment of a family Christmas, even the seemingly innocent delights of a semi-competently iced fruit cake can all too easily congeal into deep simmering resentment. Without proper mathematical planning, you can get yourself into all sorts of trouble.

'Grandad's piece is bigger than mine,' hisses your sister.

'Well, yours has more icing!' Grandad snaps back, and before long everyone is bickering, while your poor cake slowly dries out in front of them, ignored and unappreciated.

Clearly, you should have given this more thought. Dividing up a cake is a serious business, and finding a method that satisfies everyone is not so simple. If we suppose that your cake is covered in icing and filled with brandy-soaked raisins (got to keep the blood-alcohol level up somehow), then to be scrupulously fair you'll need to cut slices that each contain the same amounts of three elements: icing, raisins and the substance of the cake itself. But if the elements aren't distributed uniformly throughout, then can you really dare to dream of a Christmas free from cake-based conflict?

Luckily for you, maths has already solved this problem, and the answer is a resounding yes. It's all down to something called the Ham Sandwich Theorem.*

The Ham Sandwich Theorem says that, no matter how shoddily you construct a ham sandwich, you can always divide the ham and both pieces of bread perfectly evenly with a single straight cut. Even if your ham is lolling off the bottom piece of bread on to the work-top and the other piece of bread has fallen on the floor, there is always *some* angle at which you can hold your knife so that it will perfectly bisect all three elements in a single motion.†

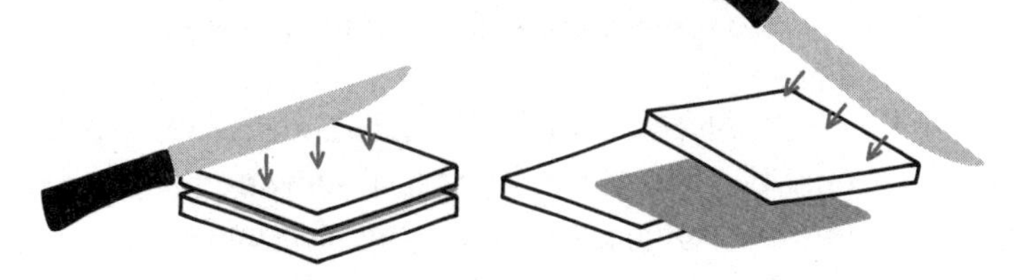

* Surely up there with the Shoes and Socks Theorem and the Hairy Ball Theorem in an unbeatable top three silliest-named mathematical theorems.

† In this example, you would probably have to slice through your work-top as well, necessitating the use of some sort of samurai sword, rather than a more conventional bread-knife, but you get the idea . . .

No, we're not suggesting that you give up on Christmas cake in favour of ham sandwiches – that wouldn't be in the spirit of things at all – but the point of the theorem is that it works for any object with three constituent parts. Including our Christmas cake.

So, even if your oven shelf is wonky and all the raisins have sunk to one end, even if your icing is so uneven it looks like you've smeared it on with a trowel, the Ham Sandwich Theorem says you can always cut your cake into two perfectly fair halves, each containing the same amounts of icing, fruit and cake. (For more on the Ham Sandwich Theorem, take a look at the note at the end of this chapter.)[1]

And, of course, if you can cut your cake into perfectly fair halves, then you can cut each of those halves into perfectly fair quarters. And you can cut those quarters into eighths, and so on, until you have enough mathematically equal pieces to feed your whole family, however large.* Problem solved.

Sorry, what was that?

You want to know how to *find* this perfect cutting angle?

Really? Ah. Well, the theorem doesn't actually tell you that. We thought just knowing that it was possible would give you the confidence to—

No?

Oh. Right. Perhaps we need to try something a bit different, then . . .

* How you deal with the spare slices is your business. You can't expect us to solve all your problems.

A CAKE-CUTTING CONUNDRUM . . .

Here's an old puzzle about cake-cutting. Suppose you have a round, uniform sponge cake and you need to cut it into eight equal pieces. Sounds easy enough? Ah, but the trouble is, you're only allowed to make three straight cuts through the cake. (Why? Who knows! Maybe you're just lazy; or in a hurry; or perhaps your knife is very fragile . . . Maths puzzles tend to be a bit light on real-world details.) You're not allowed to touch the cake either, so you can't rearrange the pieces between cuts. How do you do it?

* If you're really stumped, you'll find the answer at the end of the chapter.

There is another mathematical approach to cake-cutting that is quite a bit more practical: Austin's Moving Knife Procedure.* True, it does sound more like a horrific new form of cosmetic surgery than a

* A. K. Austin, 'Sharing a cake', *The Mathematical Gazette*, vol. 66, no. 437, 1982, pp. 212–15. In his article, Austin actually presents five different ways to share a cake fairly. We can only speculate on the mealtime injustices of his childhood that might have motivated this obsession.

solution to our dessert-division dilemma, but bear with us, because Austin's method is not only wonderfully simple, it also has the potential to liven up your Christmas dining experience considerably.

Suppose you need to share your cake among six unruly relatives.* Any one of them is liable to kick off if they don't feel they are getting their fair share – one-sixth of the cake – but, in truth, they probably don't agree on what one-sixth of the cake actually means. Your uncle likes the icing best, so he's eyeing up the part where you've plastered it on really thick. Gran prefers the boozy raisins, so she'll fight tooth and nail for a piece where they're all clumped together. Any particular slice might seem more or less valuable, not just because of its size, but also because of a person's individual preferences.

And because of these different preferences, if you just slice the cake up willy-nilly, you might end up with a piece that everyone values at less than a sixth of the whole, so whoever gets that piece is bound to get stroppy. However, with the Moving Knife Procedure, you can guarantee that everyone will end up with a slice that they are satisfied with. In fact, it's possible that most people will feel they have more than their fair share.

You can think of the method as a sort of cake auction. Here's how it goes:

* To make the explanation simpler, we'll assume that you are not having any of the cake yourself, though it wouldn't make any difference to the method if you were.

First, make a cut from the edge to the centre of the cake. We'll assume your cake is round for the minute, but it doesn't actually matter if it isn't, and it doesn't matter if you miss the centre either.

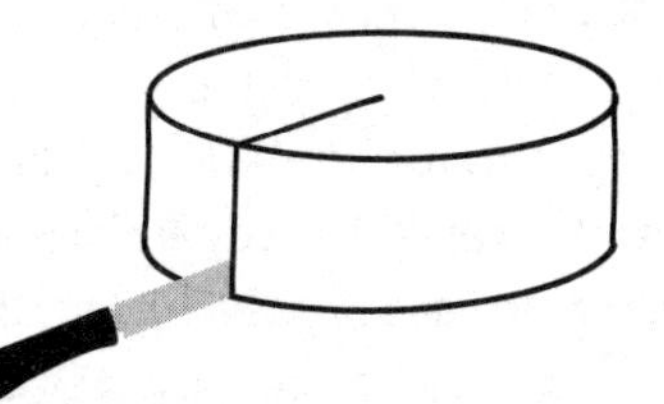

Next, keeping the point at the 'centre' of the cake, start to turn your knife slowly so that it traces out a larger and larger slice. Make sure that your whole family is gathered round attentively while you do this.

As soon as anyone thinks that the slice has grown to be worth at least their fair share (based on their own preferences), they shout 'Cut!' At this point you plunge the knife in, pass the resulting piece over to the caller, and chalk up one satisfied diner.* The most important thing here is that there is no hierarchy. Dad won't be calling dibs on the first piece, for once. Anyone can shout out at any time.

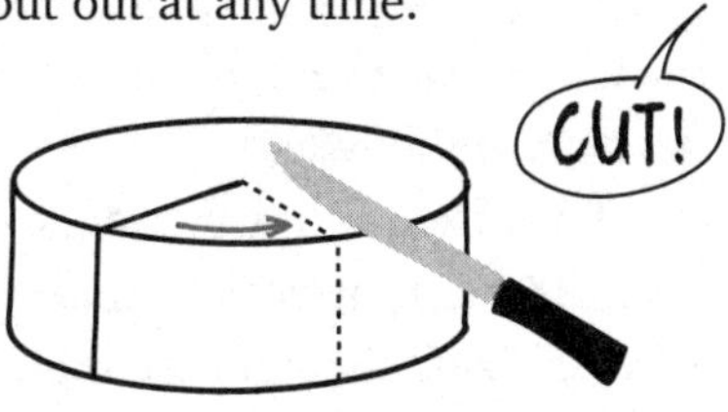

* If two people call out at the same time, since they both value the piece (and therefore also the remainder of the cake) equally, you should be able to serve either one of them without any issues.

By this point, you should already be on to a winner. Since the five people who are left have not called out yet, they must all judge that the remainder of the cake is worth *at least* five-sixths of its original value, probably more, which means there should be plenty left to satisfy everyone.

So, start turning the knife again until someone thinks that the newly traced-out slice is worth a fifth of the remaining cake, at which point they shout 'Cut!' Again, you cut the slice and pass it over, knowing that the recipient has got a piece worth at least a sixth of the original cake (by their reckoning), and the remaining four diners must judge that what is left is worth at least four-sixths of what there was to start with.

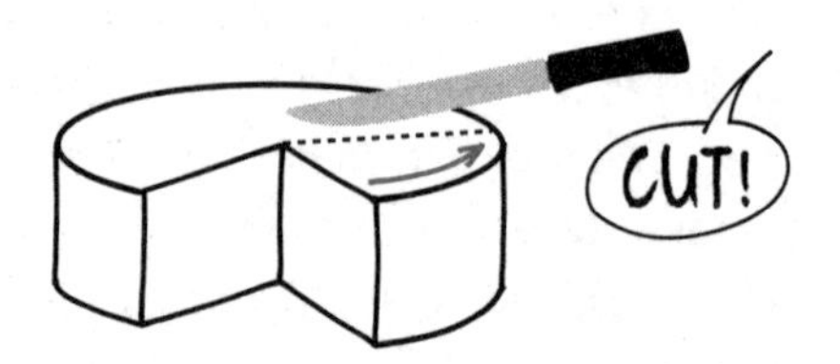

Keep going like this until there is only one person remaining. They get whatever is left of the cake, which they must value at no less than their fair share: one-sixth of the whole.

Obviously, you can adapt this method easily to suit the size of your family. If you have n family members, say, then rather than aiming for a sixth, a fifth, a quarter and so on of the remaining cake, they would be looking for fractions of $1/n$, $1/(n-1)$, $1/(n-2)$ and so on.

Austin's method is not only reassuringly mathematical, it's also, potentially, a lot of fun. You're effectively starting the Christmas games before dinner is even over – and if anyone feels hard done by at the result, they only have themselves to blame for mistiming their call.

There is one last catch to consider, though. A detail that you will only need to worry about if goodwill really is in miserably short supply among your gathered relatives.

While Austin's method is *fair*, in the sense that everyone gets a slice that is worth at least $1/n$ of the total value of the cake according to their own judgement (where n is the number of people), it falls down on another mathematical property of cake division; namely, it isn't *envy-free*.

What this means is that if the first piece goes, say, to your cousin,then while she will definitely think that her piece is a fair share (otherwise she wouldn't have called 'Cut!' when she did), there is no guarantee that she won't like the look of one of the later slices even more. She might judge that your brother has ended up with a more valuable piece than she has, for example; so if she is of a particularly envious nature, a temper tantrum might still be on the cards.

If your family is awkward enough to make you think this might be a real problem, then your best option is probably to spend Christmas alone. However, since this is not always a realistic possibility, you may have to rely on maths to pull you through again – and as it turns out, there is a mathematical approach to cake division that is guaranteed to be both fair and envy-free.* For two people, it's simple: one cuts, the other chooses, and there's no risk of either of them coveting their partner's piece. We won't lie to you, though – for three people it's a tad more complicated than Austin's method.

Here's how it works . . .

* H. Aziz and S. Mackenzie, 'A discrete and bounded envy-free cake cutting protocol for any number of agents', arXiv.org, 2016. For a comprehensible summary, see E. Klarreich, 'How to cut cake fairly and finally eat it too', *Quanta Magazine*, 6 October 2016.

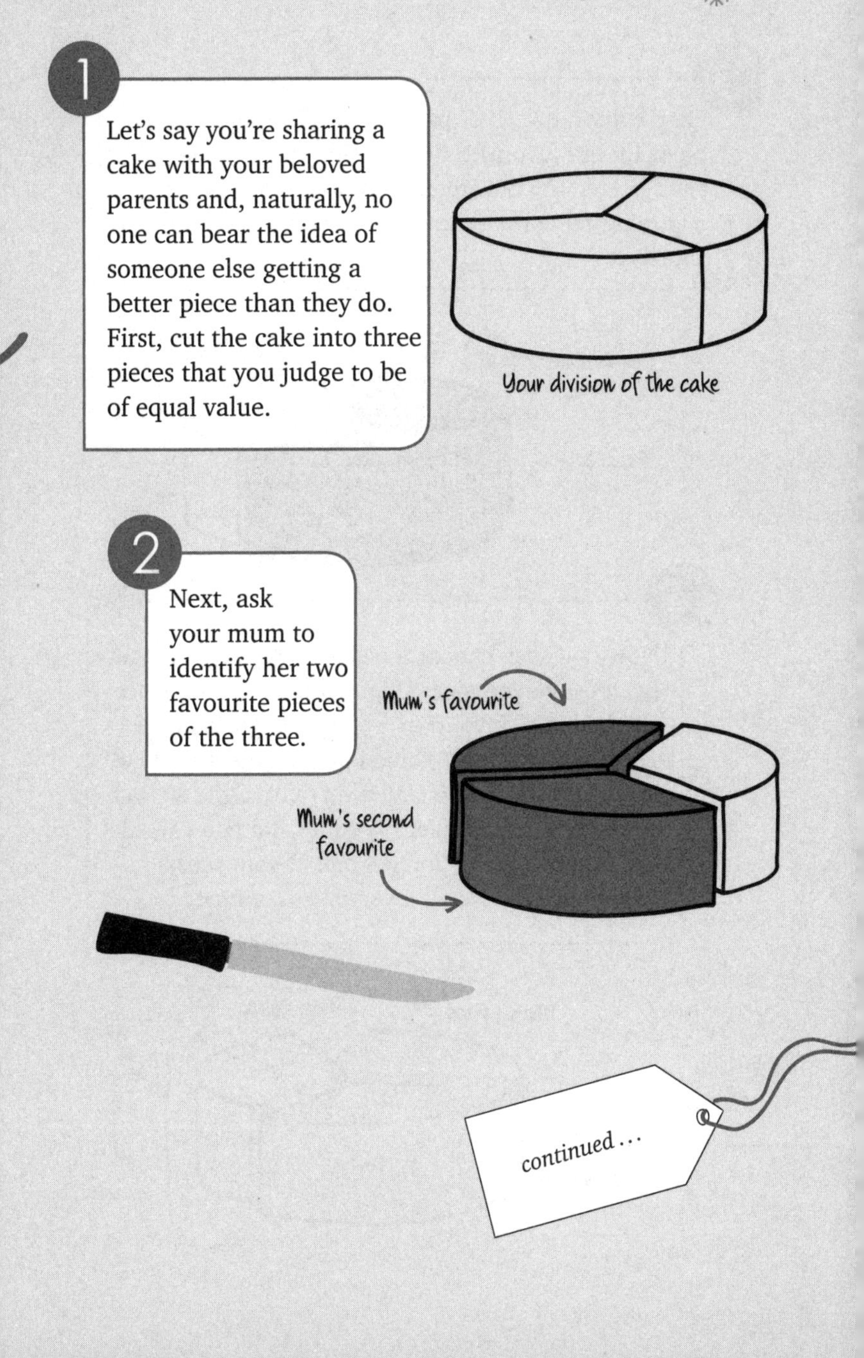
1
Let's say you're sharing a cake with your beloved parents and, naturally, no one can bear the idea of someone else getting a better piece than they do. First, cut the cake into three pieces that you judge to be of equal value.
Your division of the cake
2
Next, ask your mum to identify her two favourite pieces of the three.
Mum's favourite
Mum's second favourite
continued . . .

3

Here comes the clever part. Get your mum to trim a piece off her favourite slice, so that the two pieces she selected are now of equal value to her. Set the trimming aside for the moment; we'll come back to it shortly.

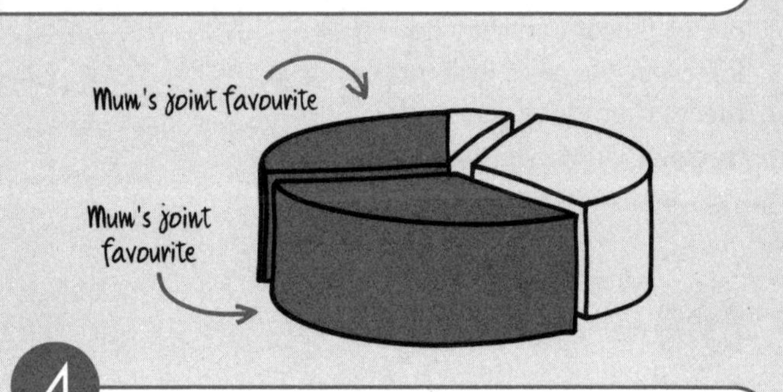

4

Now you each choose a piece. Dad first, so he gets his most valued piece. Mum goes next, and since she had joint favourites, she is also guaranteed to get one of her most-valued pieces (she must take the trimmed piece if Dad didn't). You go last, but since the trimmed piece is gone, and you valued the others equally, you get one of your most-valued pieces as well. Everyone's a winner.

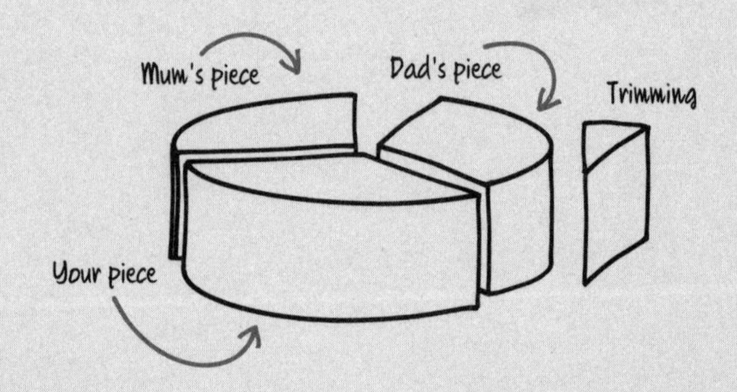

5

All that's left is to deal with the trimming. Whichever of your parents did not take the trimmed piece (Dad in this example) cuts whatever remains into three pieces that he judges to be of equal value, just like you did at the start.

6

Finally, you each choose one of these pieces. Mum chooses first, so she gets her most-valued piece again. You go next, but even if you only get your second-most-valued piece, you got to pick before your dad and you already think you've done better than your mum, since you value your original piece as equal to her trimmed piece plus *all* of the trimming. Dad goes last, but since he valued the pieces equally, he reckons he's got the best deal too.

And there you have it. Envy-free afters, guaranteed.*

For more than three people things get trickier still. In fact, a general envy-free method for dividing cake between more than three people was only discovered as recently as 2016.† You might not be too keen to try it out though, since for *n* people, the total number of cuts required could be in the region of:

Even for a family of four, this is such a mindbogglingly enormous number that it defies description in any comprehensible way – far, far exceeding the number of atoms in the universe (a paltry 10^{80} or so).‡ At one cut a second, you'll barely be getting started before the sun goes cold, and you'll

* In case you're wondering why mathematicians seem to have spent so much time thinking about cake, we probably should point out that these methods of fair division can actually be applied to all sorts of things other than baked goods. They have the potential to be used to tackle problems as diverse as dividing possessions between divorcees to resolving territorial disputes between nations. Maths is more fun if you do it while eating cake, though. That we can't deny.

† You could say it's *cutting-edge* research . . . Sorry. We just couldn't resist.

‡ J. C. Villanueva, 'How many atoms are there in the universe?', *Universe Today*, 24 December 2015.

still be squabbling over the crumbs as the observable universe collapses and dies around you.

If nothing else, it would certainly be a novel way to get out of doing the washing up.

. . . AND HOW TO CRACK IT

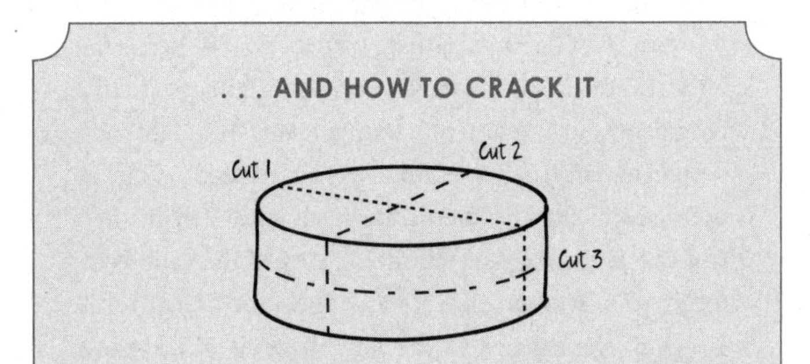

You need two cuts at right angles through the top of the cake and one, rather more unconventional, horizontal cut, parallel to the base. The key to finding the solution was to make use of all three spatial dimensions, rather than the two dimensions that standard cake-cutting is based on. In the same way, you can cut a four-dimensional cake into sixteen pieces with four cuts and a five-dimensional cake into thirty-two pieces with five cuts. We won't go into details, though, since multi-dimensional cakes are frustratingly tricky to bake.

ENDNOTE

[1] As with much of mathematics, when you're trying to understand something, it's often wise to think of the simplest case you can and build up from there.

So, let's imagine that you've got a perfectly uniform Christmas sponge cake. It's got no icing, no raisins, just plain old boring cake. It might not be particularly delicious, but if you wanted to cut it precisely in half, it should be fairly easy. Given that you can hold your knife so that 100% of the cake is to the right of it, then slowly move the knife across until 100% of the cake is to the left, there must be some point in between where precisely 50% of the cake is on one side and 50% is on the other. Plunge your knife in at exactly that point, and you've divided your cake into two equal pieces. It doesn't matter what shape the cake is or (crucially) what angle you hold your knife at, you'll always be able to split the cake fairly using this method.

Now, let's start again, but we'll make things a little trickier this time by adding in some raisins.

Our aim now is to make our cut so that each new piece has exactly half the cake and half the raisins. If the fruit is evenly spread throughout, then this is easy, so let's imagine instead that the raisins are all clumped together at one side.

This is going to require some thought. You could pick a cut that exactly bisects the cake, but leaves all the raisins in the left-hand slice. Likewise, you could swivel the knife through 180°, so that you're still splitting the cake into two equal pieces, but now all the raisins are in the right-hand slice. But hang on: if one angle has 100% of the raisins on one side, and another angle has 100% of raisins on the other, there must be some angle in between that both bisects the cake (since we've shown you can do this with any knife angle) and gives you an exact 50–50 split of the fruit as well.

This is known to mathematicians as the Pancake Theorem. (Yes, more food. Don't ask.) Adding in the third component of the pud (icing, in our case) to get to the full Ham Sandwich Theorem does make things a little trickier still, but you now get an extra dimension to play with when making your cut – you're allowed to hack through the cake at an oblique angle. Just watch your fingers.

CHAPTER 8

Christmas crackers

What is it about Christmas and weird traditions?

Take the Gävle Goat, for example. Every year, the residents of Gävle, a small city on the east coast of Sweden, celebrate the start of the Christmas season in the most natural way possible. By erecting a 10-metre-high straw goat in the centre of town.

On the face of it, this seems a rather charming ritual. A little eccentric, perhaps, but basically good wholesome fun. Sadly, though, it has also become traditional for the poor goat – despite its intended role of guiding the town into the new year – to be mercilessly targeted throughout the festive season by arsonists (and on one memorable occasion in 1976, by a hit-and-run driver). The sorry creature rarely survives through to January, and is occasionally burned down within hours of its unveiling.*

Of course, all traditions look odd from the outside. Just try explaining Christmas crackers to someone who has never come across one.† Two people struggle

* 'Festive goat up in flames again', bbc.co.uk, 27 December 2008.
† An American, for example. Crackers are largely unknown in the US. In some states they are even classed as fireworks (F. McAlpine, 'A very British Christmas; Part 3: Crackers', bbcamerica.com).

over an explosive device in a cardboard tube for the privilege of wearing a paper crown, telling a joke that everyone already knows and being accused of infidelity by a plastic fish. Ah, the true magic of Christmas!

Unless you're really taking things too seriously, the unspoken goal of a cracker-pulling session is for everyone around the Christmas dinner table to get a prize. Unfortunately, since the winner of each cracker is a fifty–fifty shot (again, unless you're taking things too seriously*), this is not necessarily a likely outcome, particularly if you have a large family. Some hasty and unsatisfying paper-hat redistribution is often necessary, with the accompanying risk of tantrums and spontaneous outbursts of violence (and that's just your grandparents).

If this scenario seems horribly familiar, you've come to the right place. Using the power of mathematics, you'll soon be able to determine the optimal cracker-pulling strategy for your family to minimize the chance that someone will be left without a gift, thus ensuring post-pudding harmony and happiness for all.†

First, we need to represent the guests around your dining table on a graph. Not the sort of graph that has lines or bars. This kind of graph is just a collection of blobs (called nodes) connected by lines

* Seriously, why are there so many articles about how to cheat at cracker-pulling (e.g. R. Gray, 'Christmas: how to cheat at pulling crackers', *Telegraph*, 19 Dec 2010)?

† Or pre-poultry harmony and happiness if your family prefers to get the cracker pulling out of the way before getting stuck in to the turkey.

(called edges). In our graph, the nodes will represent people and we'll draw an edge between two people every time they pull a cracker. We're going to assume that there is one cracker per person, so the number of nodes and edges in our graph will be the same.

Let's start by drawing a graph to represent what might happen if a family of eight – two grandparents, two parents, two children and the children's plus ones – each pull their cracker with a randomly chosen partner. Each edge has been given an arrow to point to the (randomly chosen) winner of that particular cracker pull.

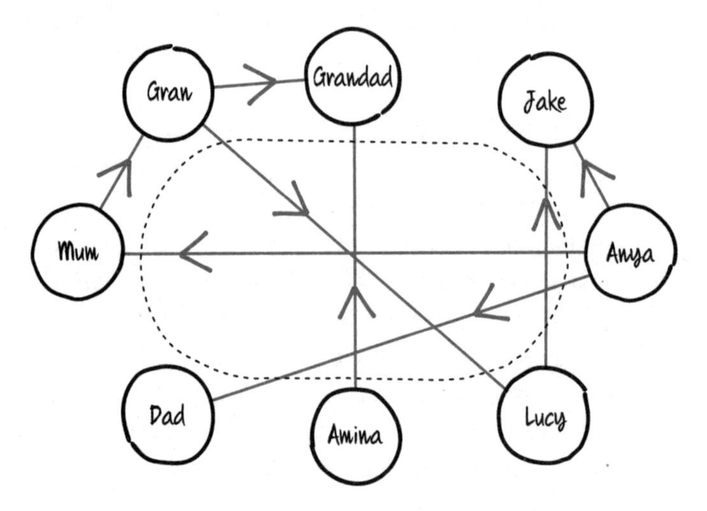

What a shambles. In this cracker-pulling fiasco, Jake has won twice, while Anya has pulled three times and not won any. This is exactly the sort of thing that happens when you recklessly try to have fun without taking the time to analyse your situation mathematically.

If this family had only thought for a moment, they might have realized that there is actually a cracker-pulling pattern that guarantees a perfect outcome – one where everyone is assured of a prize:

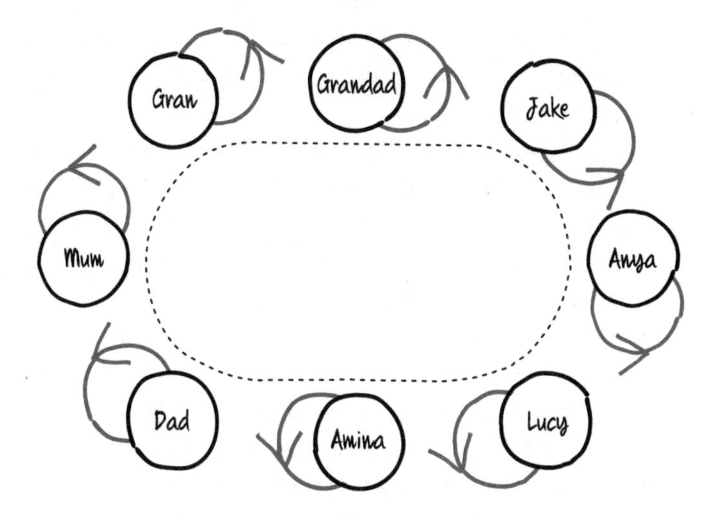

If we mathematicians had our way, this is how everyone would pull crackers. All participants are guaranteed a prize and the system works for any number of dinner guests. It also has the added bonus that no one is reaching across the table, minimizing the risk that you'll end up with a paper hat or a pack of mini-screwdrivers in the gravy.

Except . . . OK, OK. We know our suggestion is against the rules.

THE FIRST LAW OF CRACKER ETIQUETTE
THOU SHALT NOT PULL BOTH ENDS
OF THINE OWN CRACKER.

This is one of the sacred unwritten rules of crackering. Or it was until we wrote it down. Anyway, there's no way round it. Reluctantly, we're going to have to discard our perfect solution and come up with some other options.

Broadly speaking, there are two main schools of thought on cracker-pulling. First, there's the simultaneous approach, where you all cross arms, 'Auld Lang Syne' style, and everyone pulls together. Then there's the spontaneous approach, where people pair up and pull their crackers in a more haphazard asynchronous fashion.

Let's look at the simultaneous case first.

For obvious reasons, the most common pattern for simultaneous cracker-pulling is just to form one big loop around the table. Unfortunately though, if you want everyone to win a prize, this is the worst possible approach.*

To see why, let's start by picking any pair of cracker-pullers in the loop – say, Lucy and her partner, Amina. Between the two of them, it doesn't matter

* Well, strictly speaking, the worst possible approach would be something guaranteed to fail. Like two people pulling all the crackers between them. But this is the worst possible approach that still actually has some chance of success.

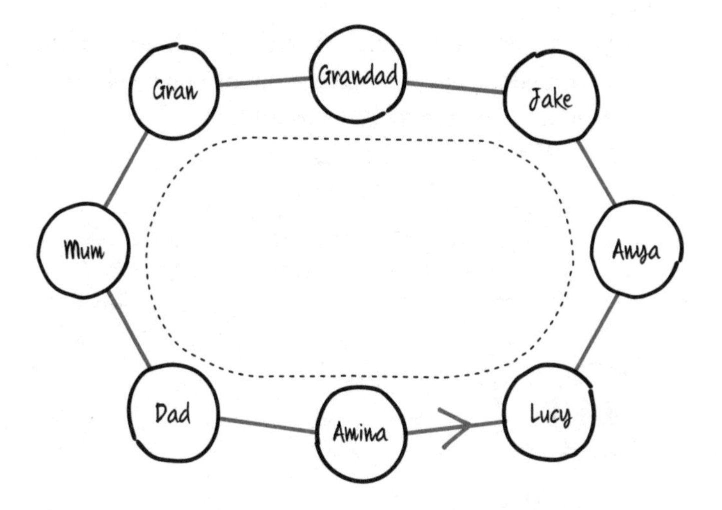

who wins, so let's say Lucy gets the prize.

But now we have a problem. Because in order for everyone to win a prize, *all the other arrows* on the graph will have to point in the same direction as Amina and Lucy's. That's seven cracker pulls that have to go our way, and since each one is a fifty–fifty shot, the probability of that happening is $(\frac{1}{2})^7$, just 1 in 128. With this method, over a century of Christmases could go by without a single successful cracker experience.

In any cracker-pulling loop of n people, the probability of everyone winning a prize is $(\frac{1}{2})^{n-1}$, because you need a favourable outcome from $n - 1$ cracker pulls. That probability gets very small very quickly as the number of people increases, but in each loop you always get that one pull that can go either way.

This is the key fact in our search for the optimal

pattern for simultaneous cracker-pulling. To give everyone the best chance of winning you need to form as many separate loops as possible.

Or to put that another way . . .

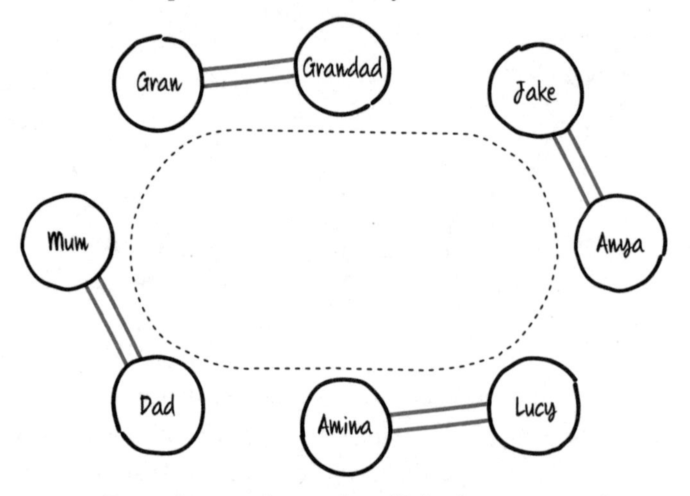

Form into pairs and pull both your crackers together (if there's an odd number of you, you'll need one group of three). Now *each pair* has a fifty–fifty chance of sharing their prizes equally, so this family's chances of universal satisfaction have jumped to $(\frac{1}{2})^4$, or 1 in 16. Much better!

In general, with this method, the probability that everyone in a family of n people wins a prize is $(\frac{1}{2})^{n/2}$ for even-sized families, and $(\frac{1}{2})^{(n+1)/2}$ for odd-sized families.

Unfortunately, we're about to get pulled up on our etiquette again . . .

THE SECOND LAW OF CRACKER ETIQUETTE
THOU SHALT NOT PULL MORE THAN ONE CRACKER WITH THE SAME PARTNER.

All right, all right. If pairs aren't allowed, then the best you can do with simultaneous pulling is to go with groups of three. You may need one group of four or five as well to make the numbers work.

Using this approach, in our family of eight – split into one loop of three and one loop of five – there would only be a 1 in 64 chance of everyone getting a prize,* still nowhere near good enough in our quest for cracker-pulling perfection.

To continue our search, let's see if we can do any better with the haphazard approach. With this method, you don't need to decide in advance who is going to pull with whom, and this makes a big difference to your chances of success. If you can check the result of each contest before deciding who should pair up next, there is a ready-made approach that will always guarantee a fifty–fifty chance of success. A good old-fashioned cracker tournament.

* For a family of n people, the probability of success is $(\frac{1}{2})^{(2n+r)/3}$, where r is the remainder when you divide the number of people by 3. That power might not initially look that sensible, but it does make good intuitive sense. Dividing n by 3 would give you the number of triangles you can split your guests into, but you also have to multiply by 2, because you need two cracker pulls to go your way in each of those triangles. Adding in the remainder makes sure the answer is a whole number and takes into account the fact that you might need one group of four or five.

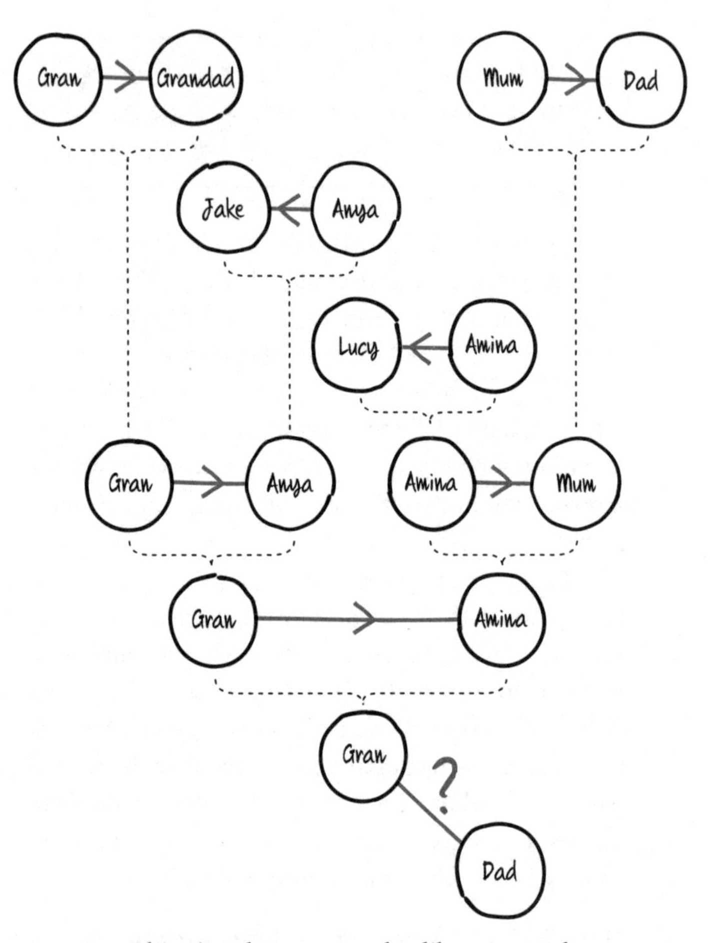

This is almost exactly like a regular sports tournament, except it's always the loser who goes through to the next round.* This ensures that every-

* Just as in a sports tournament, you may need some preliminary ties (loser-stays-on) to get the number of competitors down to a power of two (i.e. 2, 4, 8, 16 or – if you have a really huge family – 32).

one wins exactly once, except the one unlucky soul (poor old Gran, in our example) who goes through the whole tournament without winning a single pull.

But . . . there is still one cracker remaining. All our serial loser has to do is pull this cracker with one of the players who has already won and there's a fifty–fifty chance that everyone will end up with a prize.

But – uh oh. I can hear the cracker pedants clearing their throats again . . .

THE THIRD LAW OF CRACKER ETIQUETTE

THOU SHALT NOT PULL MORE CRACKERS THAN ANYONE ELSE.

Sigh. The tournament idea is out then, because Gran got to pull four crackers, and so she had four times as many turns as Grandad, who only pulled one. Classic Grandma, hogging all the fun as usual. To fulfil the third law, we're going to need everyone to pull precisely two crackers.* This means we're back to loops again – except this time we'll do it in two stages.

Let's start by pairing people off again and seeing who wins . . . (see top of next page).

* The third law of cracker etiquette effectively says that our graph must be regular, i.e. every node must have the same *degree* (the number of edges attached to it). Since there are the same number of nodes and edges (everyone has a cracker) and every edge has two ends, every node must have degree two.

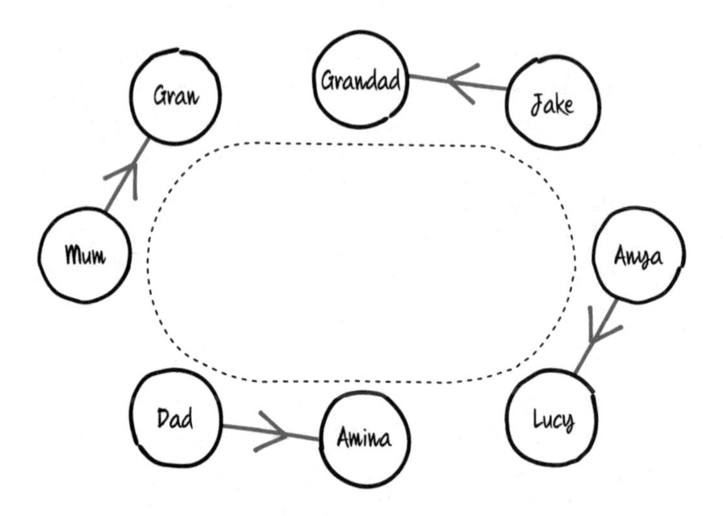

Now, if we add the remaining crackers to create one big crazy loop, and we make sure to arrange it so that it incorporates the arrows we already have, all pointing in the same direction round the loop, then we'll obey the third law *and* we'll only need the remaining pulls to go in our favour. That means we're only sweating on half of our total number of cracker pulls if there's an even number of people, or just over half otherwise (see diagram opposite).

With this two-stage method, our family's chances of success are $(\frac{1}{2})^4$, or 1 in 16. For a family of size n the probabilities are $(\frac{1}{2})^{n/2}$ for even-sized families and $(\frac{1}{2})^{(n+1)/2}$ for odd-sized families, the same as in our earlier (illegal) pairs method.

We do hope this cracker-pulling advice brightens up your Christmas. Whatever happens, though, do spare a thought for the 1,081 people who gathered

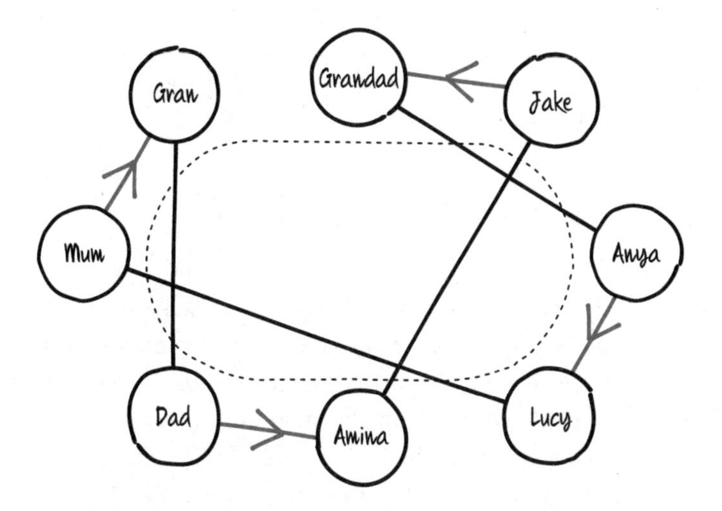

at the Harrodian School in Barnes in December 2015. Although they successfully broke the world record for the longest Christmas cracker-pulling chain in history,* we can be pretty sure that a lot of them went home empty-handed. Since they were using the simultaneous loop method,† the probability of everyone involved winning a prize was just $(\frac{1}{2})^{1080}$, an *indescribably* minuscule chance of about 1 in ten septencentillion.‡

Just imagine how disappointed they must have been . . .

* Source: guinnessworldrecords.com.
† Assuming they joined up the ends of the chain to form a loop.
‡ Ten septencentillion is a 1 followed by three hundred and twenty-five 0s, a number so large that calculating it required a special piece of software (speedcrunch.org) and even finding out its name required a trip to the local library (J. H. Conway and R. K. Guy, *The Book of Numbers* (1996), pp. 15–16).

CHAPTER 9
The Queen

So, Christmas dinner is over. Crackers have been pulled, jokes have been told and some poor soul has been abandoned in the kitchen to spend the rest of the day washing up the mountain of greasy dishes that's teetering in the sink. For everyone else, it's time to collapse into the sofa, count your Christmas presents and try to avoid eye contact with your most obnoxious relatives for the next five hours or so.

But then, just as you're getting comfortable, someone turns on the TV and your lazy afternoon is interrupted by the opening bars of the national anthem. Whether you like it or not, the Queen is crashing your Christmas.

Since she came to the throne in 1952, the Queen has given well over 60 Christmas broadcasts, only failing to appear in 1969, when she was presumably still recovering from Woodstock or from helping Nixon to fake the moon landing. Britain couldn't seem to get through the bleak midwinter without her, though, since an avalanche of disappointed fan mail prompted her to issue a written message as a stopgap, reassuring everyone that she would be making a TV comeback the following year. It seems she's never again managed to think up a good enough excuse to get out of it, and her Christmas message has been a reliable feature of the big day ever since.

In total, including that half-hearted 1969 effort, the Queen has delivered over 42,000 words of festive musings.* That is roughly equivalent to two plays by William Shakespeare† or approaching 400 renditions of 'Rudolph the Red Nosed Reindeer' (and what a Christmas tradition that would have been . . .). Alternatively, based on the official texts of her speeches, she could have issued the messages in over 1,700 tweets – about 27 per year – though this would have been a pretty unorthodox approach to communicating with her subjects back in the 1950s.

Across all her Christmas messages, the Queen has used over 4,300 different words.‡ If you've ever wondered how her vocabulary stacks up against the luminaries of the world of hip-hop (and let's be honest, who hasn't?), then wonder no more. The 3,991 distinct words that appear in the first 35,000 words of her Christmas messages put her behind 79% of the world's most famous hip-hoppers, in line with Snoop Dogg (3,974) and Kanye West (3,982), but behind such noted wordsmiths as Nicki Minaj

* The texts of all the Queen's Christmas messages are available to view at royal.uk.

† *Julius Caesar* and *Richard II* total 42,126 words, according to opensourceshakespeare.org.

‡ An exact figure is actually not easy to pin down. It depends how you choose to count awkward things like numbers and a variety of pesky hyphenated words. This caveat also applies to the figure of 3,991 that is mentioned later.

(4,162) and Jay-Z (4,506), while she is positively left for dust by the vocal gymnastics of the Wu-Tang Clan (5,895).* Shakespeare, for the record, also beats her comfortably with a score of 5,170, though he did make a lot of words up.

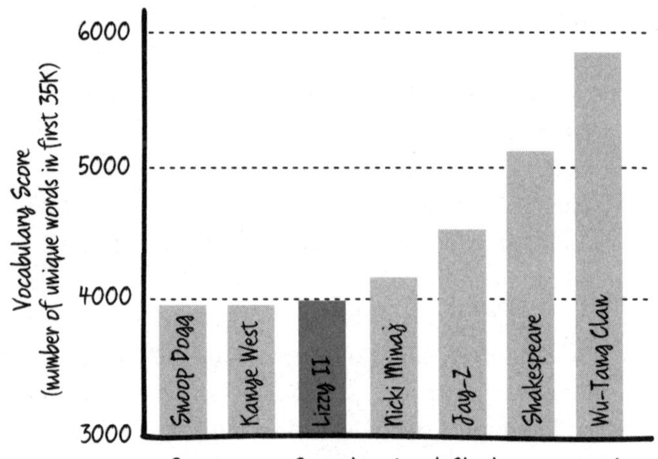

As with us peasants, Her Majesty's most used words are pretty boring – 'the' (over 2,500 occurrences), 'of' and 'and' (over 1,800) are the most common – but if we just focus on more interesting words – say, the nouns – we can glean a little bit more about what Christmas really means to our Liz:

* The vocabulary counts for hip-hop artists (and Shakespeare) were calculated by Matt Daniels ('The largest vocabulary in hip hop', *Polygraph*). 35,000 words is the number examined in that investigation.

TOP 25 NOUNS USED IN THE QUEEN'S CHRISTMAS MESSAGES*

1	year/years	297
2	Christmas/Christmases	291
3	people/peoples	244
4	world/worlds	200
5	Commonwealth	174
6	family/families	165
7	time/times	157
8	life/lives	151
9	child/children	108
10	country/countries	103
11	other/others	101
12	day/days	96
13	man/men	94
14	message/messages	77
15	future	68
16 =	nation/nations	65
16 =	peace	65
16 =	way/ways	65
19 =	part/parts	64
19 =	service/services	64
21	home/homes	58
22	community/communities	57
23	hope/hopes	55
24 =	faith/faiths	52
24 =	God	52

* Only uses of 'other' as a noun (e.g. 'each other') were counted, though it was used many more times as an adjective (e.g. 'other people'). Uses of 'hope' as a verb were similarly excluded. The words 'lives', 'hopes' and 'future' were overwhelmingly used as nouns, rather than as verbs or adjectives.

There are words about time, words about society, some religious words . . . All very good and noble, of course, but also just a teeny tiny bit dull.

What we really need is some way to liven things up. And given that Her Maj is known to enjoy a day at the races from time to time, what better and more respectful way to do that than by introducing a bit of good honest recreational gambling.

And so, ladies and gentlemen, in that spirit, we bring you . . . Queenie Bingo!

Yes, that's right. We've created four probabilistically balanced bingo cards to brighten up your Christmas afternoon. Divvy them up,* gather round the telly, and the first to cross off a complete row, column or diagonal is the winner. We guarantee you'll never have concentrated harder on what good old Lizzie has to say.

The words are arranged so that every winning line contains one common word, one uncommon word and one rare word,† but the Queen is so predictable that every card should be a winner eventually. And if you do fancy a flutter, why not optimize your day still further by wagering your unwanted gifts on the outcome. You could even drape a ribbon across the door to the airing cupboard and let the victor cut through it for the full queenly experience. Festive fun for all the family.

* You'll probably want to buy three more copies of the book at this point, of course . . . No? Oh, go on then, you can copy the cards if you want.

† Roughly speaking, 'common' words have been used between two and five times per year (on average), 'uncommon' words between once and twice a year, and 'rare' words less than once a year.

CARD ONE

	year years	faith faiths	together
new	love / loves loved	world worlds	
Commonwealth		country countries	today
help / helps helped	hope / hopes hoped		life lives

CARD TWO

time times		home homes	day days
good	God		Christmas Christmases
	world worlds	great	wish / wishes wished
peace	country countries	Commonwealth	

CARD THREE

nation nations	family families		child children
man men		time times	work / works worked
Christmas Christmases	war wars	future futures	
	great	woman women	people peoples

CARD FOUR

new	people peoples	courage	
	service services	family families	child children
life lives	happy		community communities
old		message messages	year years

However, before we get too carried away with all this frivolity, let's not forget there's a human side to this tradition as well.

Having to come up with a suitably seasonal speech every year can't be easy. If only we could use the power of mathematics to help her out a bit . . .

With over 60 speeches already written, there's a lot of raw material to work with. Perhaps we could come up with some sort of mathematical algorithm – a step-by-step process – that would use those old messages to write a new speech in the same style, leaving our noble nonagenarian free to kick back and enjoy her first stress-free December since 1951.

The simplest way to generate a new speech would be to pick words at random from the combined texts of all the Queen's previous Christmas messages, and smoosh them together. Using this method, the frequencies of the different words should be about the same as in previous speeches, and if we treat each full stop as a separate word, the sentences will be about the right length too.*

However . . . Well, here's an example of a sentence produced by this method. You can judge for yourself:

'New without and Iraq this who in years we lives it been united their the a by come God and.'

* For simplicity's sake, we can forget about all other punctuation, except for things that are intrinsic to words, like apostrophes and hyphens.

I think we can agree that this lacks the necessary dignity for a regal address. It looks like we'll need a slightly more sophisticated approach.

This is where the maths comes in. Rather than pick every word at random from all the previous Christmas messages, we can use a cunning little tool called a Markov chain.

A Markov chain is a special kind of sequence. Since we're writing a speech, we're interested in sequences of words, but the sequence could just as easily be numbers or colours or Von Trapp children. The maths is the same.

As in our previous effort, this method will create a new speech from what the Queen has said before. However, what's special about a Markov chain is that the word you write next depends on the word you've just written. For example, suppose we've just written the word 'shining'. The Queen has used that word four times in total, twice followed by 'example', once by 'examples' and once by 'tree'. The Markov chain will randomly select one of these three words to write next, according to their frequencies, choosing 'example' with a probability of 50%, and 'examples' or 'tree' with probabilities of 25%.

We also need to know how to choose the very first word of our speech, but to do that we can just pick one of the previous speeches at random and take the first word from there.

One easy way to understand how a Markov chain works is by using a directed graph, like the ones in Chapter 8. You can see how this works on the next page.

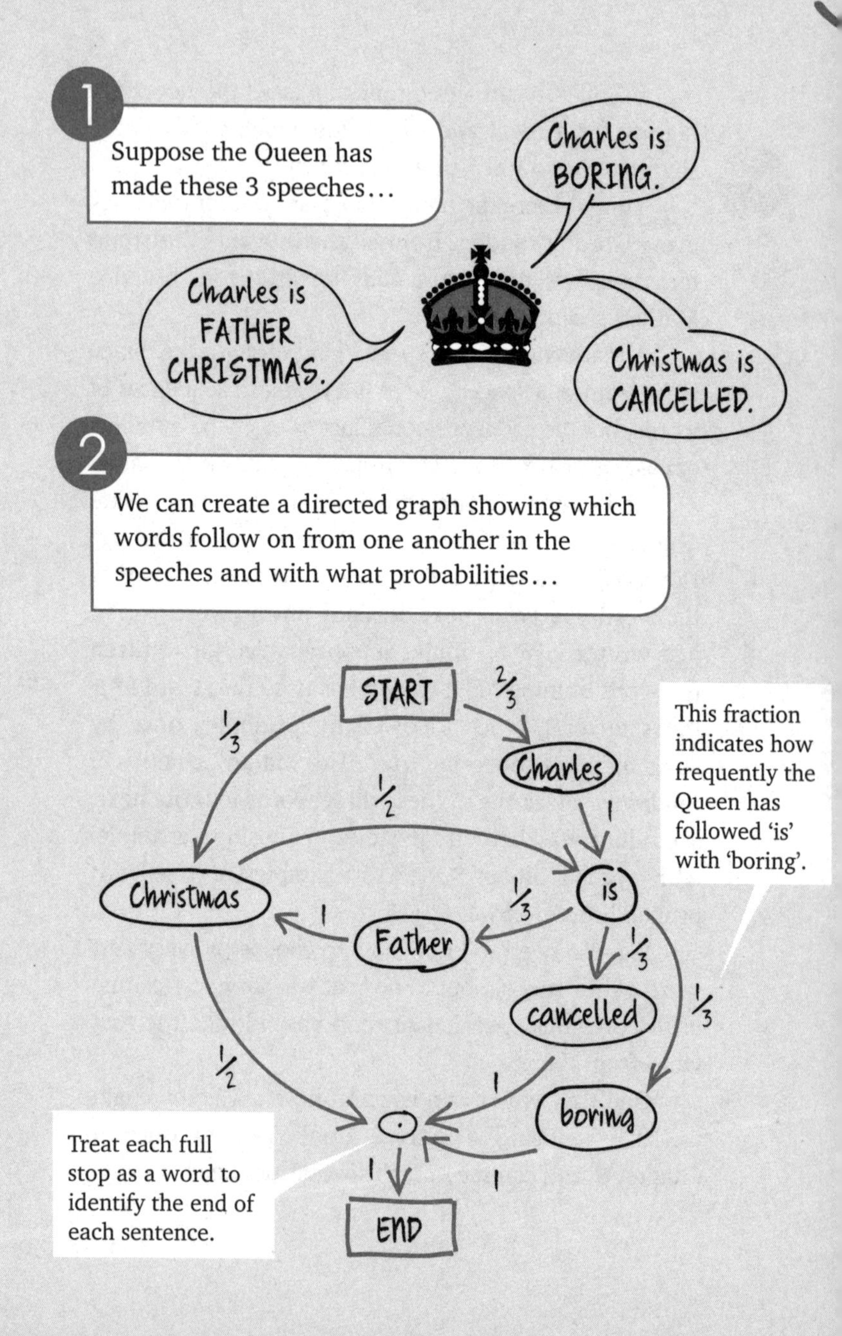
1
Suppose the Queen has made these 3 speeches…
Charles is BORING.
Charles is FATHER CHRISTMAS.
Christmas is CANCELLED.
2
We can create a directed graph showing which words follow on from one another in the speeches and with what probabilities…
START
2/3
1/3
Charles
1
1/2
This fraction indicates how frequently the Queen has followed 'is' with 'boring'.
Christmas
is
1/3
1
Father
1/3
cancelled
1/3
1/2
1
boring
.
1
1
Treat each full stop as a word to identify the end of each sentence.
END

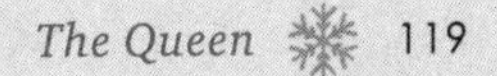

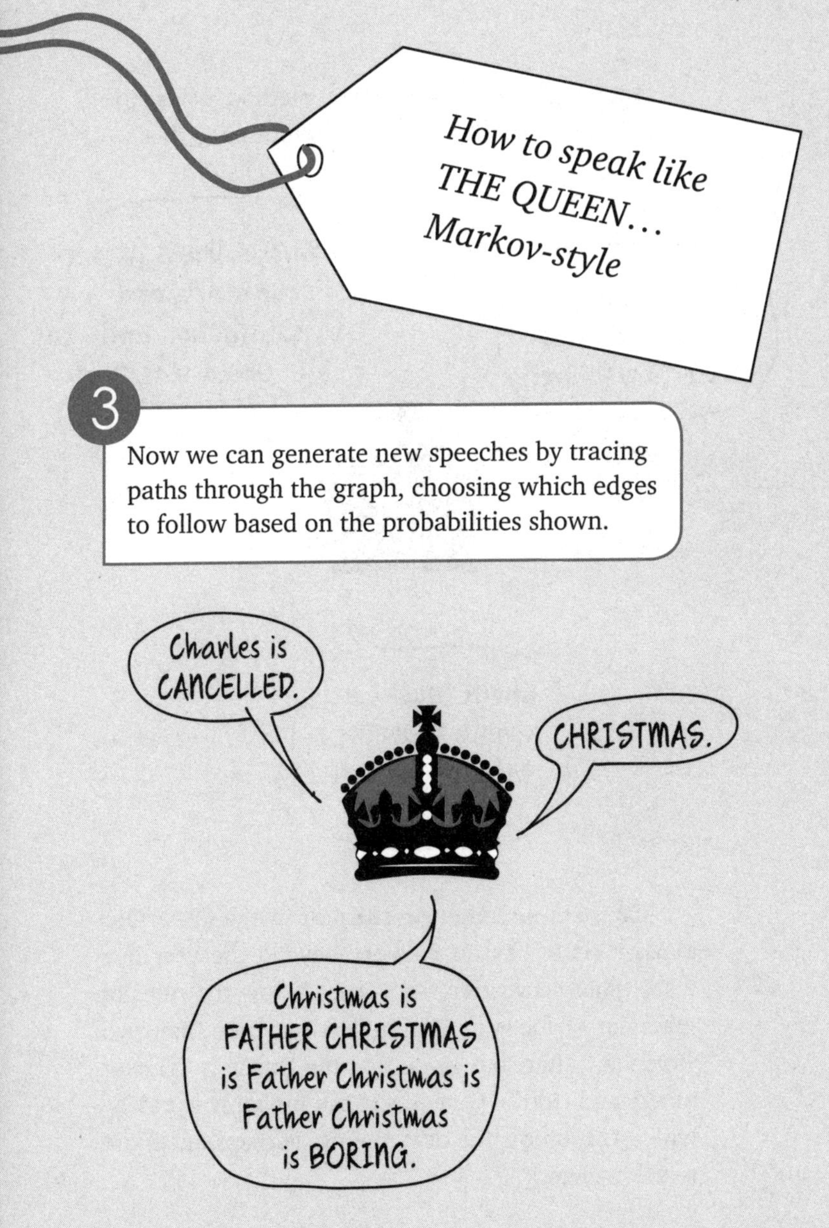
How to speak like THE QUEEN… Markov-style
3
Now we can generate new speeches by tracing paths through the graph, choosing which edges to follow based on the probabilities shown.
Charles is CANCELLED.
CHRISTMAS.
Christmas is FATHER CHRISTMAS is Father Christmas is Father Christmas is BORING.

Applying the Markov chain method using the Queen's actual speeches produces output like this:

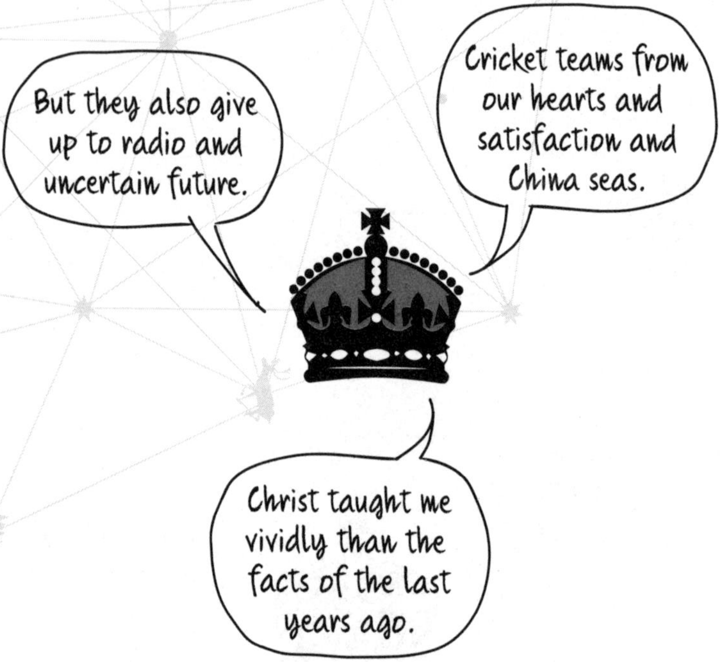

OK, so these sentences still don't make sense. Our Markov chain has no memory beyond the very last word that was written, so it can't follow the rules of grammar or focus on any subject for more than two words at a time, but given that the Queen is (a) over ninety and (b) the Queen, she can probably get away with a certain degree of rambling. Perhaps one more tweak will do it.

We can make our Markov chain a little more sophisticated by letting it remember more than one word at a time. A two-step Markov chain, for example, would consider the last two words that were written and look at what words came next over all the previous times the Queen used those two words together. For example, the Queen has used the phrase 'I look' five times, followed twice by 'back', twice by 'forward' and once by 'to'. This means that if we have just written 'I look', the Markov chain will select 'forward' or 'back' to be the next word with a probability of 40% each, and 'to' with a probability of 20%.

It turns out that if we use any more than two steps, our Markov chain starts to reproduce large chunks of the original speeches, which isn't what we're after at all. We don't want the Queen to be accused of recycling her old material. Two steps is the sweet spot.

Without further ado, then, here is our Christmas gift to Her Majesty. A ready-made speech, hot off the press from our two-step Markov chain, mathematically guaranteed to be stylistically indistinguishable from all the others that she has given.*

* Loyal subjects that we are, we've added some punctuation and paragraphing to make the speech easier to read.

It is with us.

If, as its head, I can recall only the gloomy side of life. What a wonderfully exciting prospect and perhaps reflection on the differences between the wealth of nations as I possibly can.

The birth of Christ. It is not indifference, but is rather a special family; a country where your children can enjoy the festival of Christmas. It showed that what people are suffering the pain and grief. Our children at our Christmas party, the meeting also showed that what they do is unseen and unrewarded. The context of the brotherhood of man.

May tolerance and respecting others. He bore this with all its origin in the direst of circumstances. They came to frightened shepherds with hope. We know the arrival of our Commonwealth family of nations as we were helped to understand.

It is that every religion has something to my own people and one for people with their families; Christmas will not have been entrusted with something of their saviour, but it is the best possible upbringing for their steadfast loyalty to one's favourite football club to be recognized for service to the weak and innocent victims. They have sold their heating systems even in our hearts, was that we can also help in this tragic situation.

> The upward course of history and ready and willing to support young people and their kind, who by resisting the bully and the peace movement in Northern Ireland. All those who devote their lives.
>
> I was in the last twelve months, which has meant so much. But while bravery and service to their relations and even families. It is perhaps in this tragic situation. The dedication of the responsibility for the real peace on Earth – goodwill toward men, which is known to many. Like many others, was much heartened by the need to go there.
>
> It is sobering and inspiring to remember, as I am sure that in spite of all.

Yes, it may make her sound like she's overdone it a little on the sherry, but hey – isn't that what Christmas is all about?

Anyway, we're sending this speech straight round to Buckingham Palace. So make sure you have this page open at three o'clock on Christmas afternoon and you'll be able to read along with our beloved sovereign.

That's got to be enough to get our names on the New Year's honours list, right?

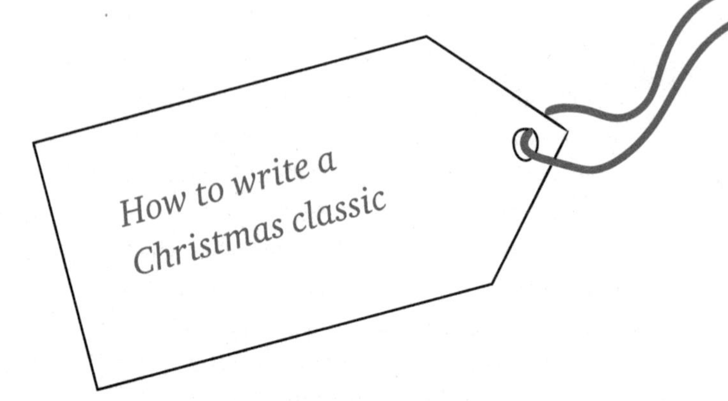

Of course, the method we've used to generate the Queen's Christmas message could have all sorts of other yuletide applications too. Just imagine how much time you could save if you used a Markov chain to write your thank-you letters, for instance (and how unlikely it would be that you would ever again receive presents from the people you wrote to).

Alternatively, you could use the technique to try to write the ultimate Christmas song. You can see our effort opposite. It was distilled from the lyrics of 80 seasonal favourites using a one-step Markov chain, and should be sung to the tune of . . . well, pretty much whatever you like, really, so long as you throw in some sleigh bells.

The Happiest Christmas We Believe . . .

VERSE 1

O Christmas time windows are brightly.
There'll be seen is that mother mild where,
You better not – Here is the jingle horse,
Was born, is falling and everyone know.

CHORUS

It's Christmas dinner, a glorious morn!
In a pear tree thy candles shine,
Out in a winter wonderland.
I went together: 'Fa la la la la la
la la la la la la la la la la!'

VERSE 2

On, it's Christmastime, there's no ear,
May ye shepherds quake at turkeys;
Ducks and a-mingle in a mystery with bells,
Swing and fill us now and joy!

VERSE 3

We'll have snow on Christmas Day.
Good girl, Santa must be!
Now the angels singing in our way.
'Ding dong, ding dong, ding dong,
ding dong, ding dong!'

CHAPTER 10

How to win at Monopoly

Once the Queen has delivered her festive message (hopefully a little more coherently than we suggested in Chapter 9) it's time to stoke the living-room fire, settle into a comfy chair with a glass of Baileys and relax, ready for the most magical of all Christmas traditions: a full-blown family argument.

You might manage to navigate the most obvious pitfalls – the heated political discussions with Grandad, your parents' imprudent prying into your love life – but there's still no escape. Because if there is one thing that is guaranteed to turn a nice afternoon of family bonding into a major shouting match, it's a good old-fashioned game of Monopoly.

Officially, Monopoly ends when all players but one go bankrupt. In reality, it ends when your sister accuses one or all of you of cheating, flips the board across the room and storms off in a shower of miniature plastic houses. Monopoly, whose sole objective is to have you force your friends and family into poverty, seems, above any other board game, to have the ability to turn even the most angelic of us into a monster. But if you are going to be involved in a vicious Christmas Day fight with your family, you might as well emerge the victor.

The key to success in Monopoly is noticing that not all properties are created equal. This is where mathematics steps in to help. Knowing which sets are likely to give you better returns is your ticket to board-game glory, ensuring it's you with the smug, satisfied smile on your face come Christmas evening while your relatives weep into their trifle bowls over their devastating defeat.

Calculating which are the best sets to go for all comes down to a question of probability. First you'll want to know which squares are most likely to be landed on, then we can use that to explore which are the most profitable.

★ STARTING AT GO

Let's begin, as the game does, with your first turn, starting at Go. With your two dice you could throw anything from 2 (double 1) to 12 (double 6). Some totals are more likely than others since there are more combinations of dice rolls that will produce that result. For instance, there are three different ways to throw a 4, but only one way to throw a 2, so you'll be three times more likely to land on Income Tax as on Community Chest.

Overall, based solely on dice rolls, the probability of where you'll end up after your first turn is as shown on the next page, with a peak at seven squares away from Go:

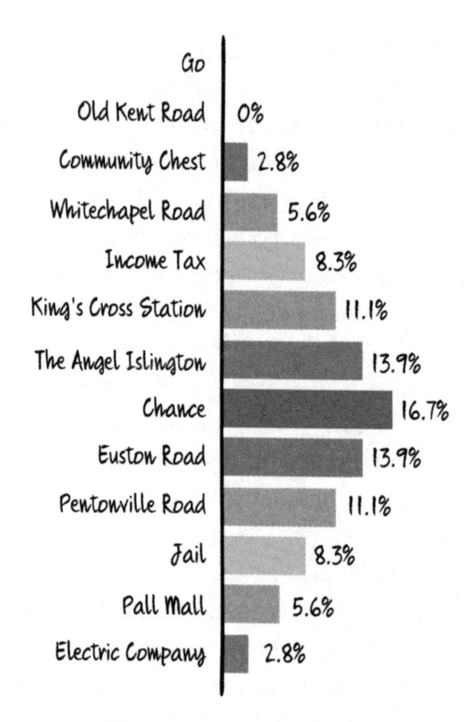

But Monopoly would be very boring indeed if the outcome depended entirely on dice throws.* In fact, there are a few curveballs in the game that mix things up and alter these probabilities.

One such curveball is Community Chest. Most

* If dice rolls were the only thing that dictated where you ended up, you would be equally likely to land on any square over the course of an entire game. Each time you moved to a new square this pattern of dice rolls would stretch out over the 12 places ahead of you, meaning that every square on the board would appear in the dice-roll pattern somewhere for the 12 preceding squares. As you travelled around the board again and again, the total probabilities of landing on each square would average out to equality.

of the 16 cards will just ask you to pay a fine or will compliment you on your radiant looks by awarding you second prize in a beauty contest. But there are three cards that will send you somewhere else on the board:

So, since your first throw gives you a 2.8% chance of landing on Community Chest, 13 out of 16 times you'll stay there, but you also have a 2.8% × 1/16 = 0.175% chance of drawing the card that sends you back to Go.

Likewise, there's a 0.175% chance that your first throw will lead you to pick the Community Chest card that sends you to Old Kent Road, and 0.175% of the time you'll end up with the worst start to a game imaginable by pulling the card that sends you directly to Jail.

The Chance cards are even more mischievous. Of the 16 Chance cards, seven will send you to different destinations (one of them being Jail again).*

* The others are Go, Income Tax (three spaces backwards), Pall Mall, Marylebone Station, Trafalgar Square and potentially the most costly card of all: Advance to Mayfair.

Factoring in all of these cards, the probabilities for where you're likely to end up on your first turn are as shown here.

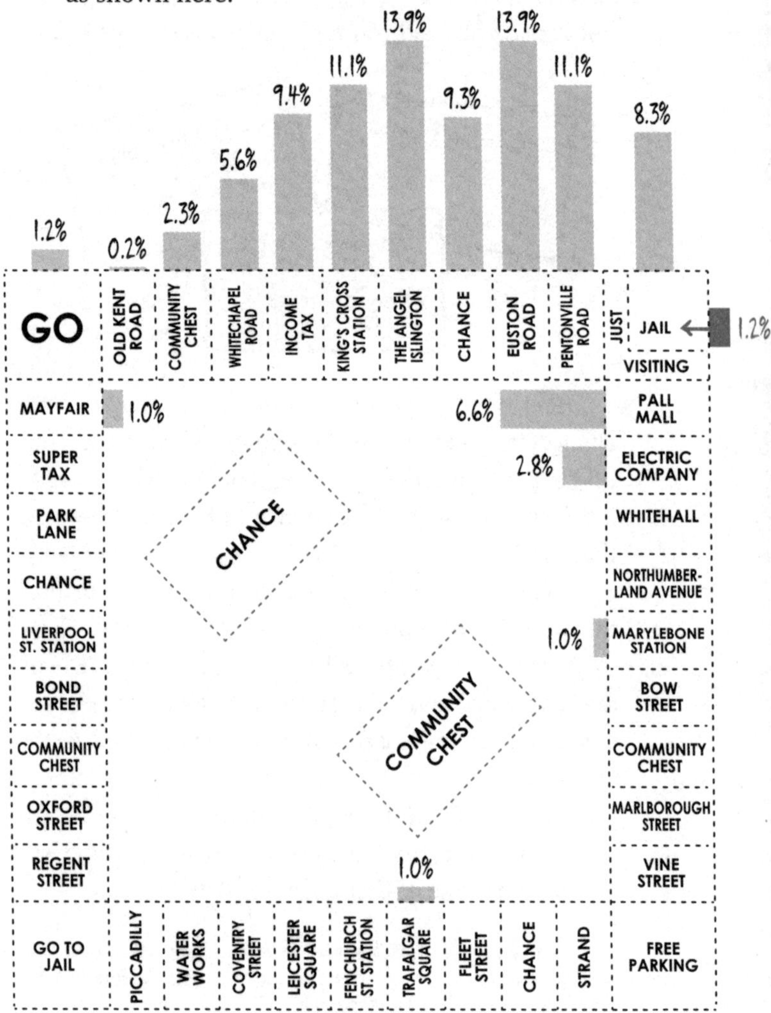

In total then, with the cards properly shuffled, you've got a 1.2% chance of ending up in Jail on your very first go (as opposed to an 8.3% chance of just visiting it). Assuming you can only stomach one round of Monopoly a year (surely a fair assumption), this means you should expect to start your Christmas evening behind bars about once in a lifetime.*

So far, these probabilities only apply to your very first throw of the game. We now have to repeat this process, calculating the probabilities of getting to and from every single square on the board, while taking into account the additional pitfall of going to Jail after throwing three consecutive doubles.[1]

Right now, you're probably eagerly sharpening your pencils and gathering your family around you for an evening of communal calculation. And what better way to while away those cold winter nights than a session of good old-fashioned arithmetic, preferably accompanied by a few rousing choruses of 'Ding Dong Merrily on High'.

However, for those of you who fail to see the joy in abandoning your mince pies and running gleefully to your spreadsheet, we've done all the hard work for you. We've wrapped all of these probabilities into a huge table, known as a *Markov matrix*, that will tell

* With one game a year and a 1.2% chance, you should expect to head straight to Jail about once every 83 years. Based on data from 2015, UK life expectancy is just short of this, at 81.2 years (World Health Organization, *World Health Statistics 2016: Monitoring health for the SDGs* (2016), Annex B: Tables of health statistics by country, WHO region and globally).

you the chances of landing on any square, from any other square. It looks just like those mileage charts you get in the back of road atlases,* except with probabilities instead of distances.

This Markov matrix can tell you, given a start position, where you're likely to end up on your next move – but it doesn't yet take into account the fact that some start positions are going to be more likely than others, and that's an important extra piece in the puzzle if we want to know the overall probabilities of landing on each of the properties throughout the entire game.

For instance, since there are a number of different ways to end up in Jail, it will be the most visited square on the board. That means that any squares within six, seven or eight of Jail are also going to be visited a lot (looking at you, Bow Street), and so too are squares that are within six, seven or eight of there.

OVERALL CHANCES OF LANDING ON A PROPERTY

But hang on, we hear you cry. Probabilities? Some chap called Markov? This is all beginning to sound mightily familiar.

Before you refer back to your notes† from Chapter 9, you're quite right to notice the similarity, because it's the very same sort of Markov chain that will help us work out the probabilities we seek. In Chapter 9, it helped identify

* For anyone under 25, a road atlas is what people used before Google Maps. You didn't miss out. They were a bloody nightmare.
† You *have* been taking notes, right? What did you think this book was? A frivolous stocking filler?

the word you write next based on the words you've just written. Here it reveals where you'll end up on the board depending on where you've just been.

You might think we're cheating a bit, using the same maths for two completely different bits of Christmas, but actually this connection precisely illustrates one of the genuinely wonderful things about applied mathematics. Often you'll discover seemingly unconnected areas of the real world – like royal speeches and tedious board games – that are in fact underpinned by startlingly similar mathematical principles. It's like building a bridge between two parts of the real world. Suddenly – with that mathematical connection – everything you know about one area applies to the other.

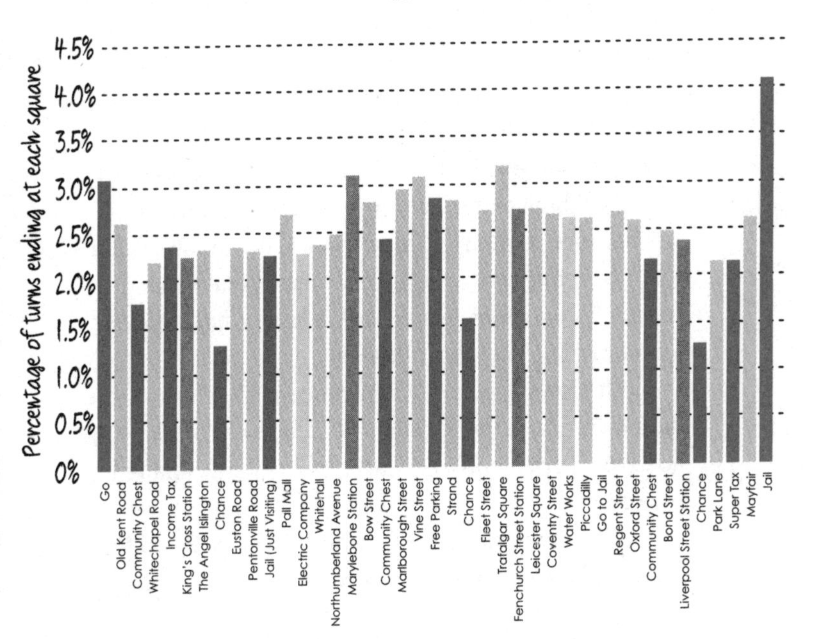

Since you've met Markov chains before, we won't bore you with the details of how to use the matrix to work out the full range of probabilities, but if you'd like to know a little more you can find an explanation in the notes at the end of the chapter.[2]

Instead, let's look at the results shown on the previous page: the total chances of a player ending up on any square at any point in the game.

As there are a number of ways to get there, Jail is by far the most visited square on the board. The Go to Jail square, by contrast, is not 'visited' by anyone, since no one who lands on it ends up staying there.

The orange property set benefits from all the ex-cons leaving their cells, and after their next turn the reformed criminals will likely end up somewhere between the reds and yellows, hence the popularity of those sets. Trafalgar Square, with its own dedicated Chance card directing people to it, gets an extra boost, making it the second most visited square on the board.

The Chance and Community Chest squares score low because so many of the people who land on them end up being sent elsewhere. The property that is visited least frequently is Park Lane, where players spend just 2.1% of their time.

But don't give up on Park Lane just yet. Even though landing there might be relatively unlikely, a hotel on Park Lane can earn you a game-changing £1,500.

Ultimately, Monopoly is about money. So, to find the best strategy, we also need to take into account how much you can expect each property, fully loaded with hotels, to earn for every roll of the dice and how much investment is required to get you there.

SO WHICH ARE THE BEST PROPERTIES?

To make a really well-informed decision, you're going to want to know how quickly you'll make your money back and how quickly you'll earn the big bucks once you do.

This graph shows a selection of properties from each set. With a hotel on each, your earnings are shown up the vertical (y) axis and the total number of your competitors' rolls along the horizontal (x) axis. Each

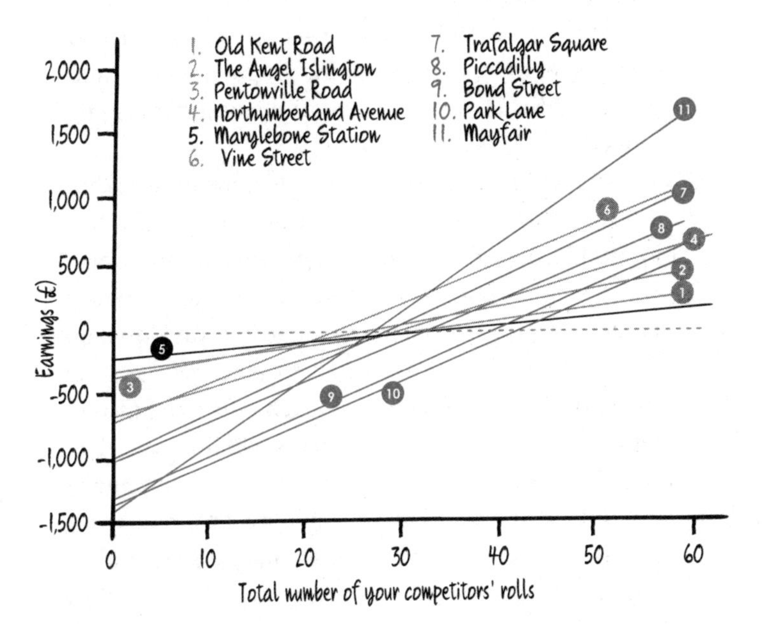

property starts off with a negative value, as you have to invest in them to begin with; as soon as a property crosses the dotted line at zero the amount it's earned in rent exceeds the amount it cost you. Anything else it earns is pure profit.

Of the properties you can build on, those from the brown and light blue sets start closest to the dotted line because they are so cheap to buy and build on. The slope of Old Kent Road on line 1 is very shallow, meaning it takes a long time before it will start making money and even when it does you're never going to make much. The Angel Islington on line 2 does a lot better and is up there with the best earning sets until about 25 rolls.

Vine Street (line 6), with its steep slope of £30 expected earnings per roll on the graph, is quickly into profit and holds on to the lead of the best performing property until well over 30 throws.

In Monopolyland, unlike real life, the railways are at least relatively affordable. However, very much like in real life, it's unlikely they will get you anywhere fast.

Mayfair (line 11), however, is the standout property on the board. Despite starting way down on the y axis, it soon streaks into first place, and it stays there for the rest of time. If you can afford to buy and build on Mayfair, it looks like it's a winning strategy.

But Monopoly doesn't let you build on single properties. If you plan on building any houses or hotels, you've got to invest in complete sets. And so this graph – although it demonstrates why so many people go for Mayfair – doesn't quite tell the whole story. For that, you've got to apply the same idea to the performance of complete sets.

WHICH SETS TO SET YOUR EYE ON

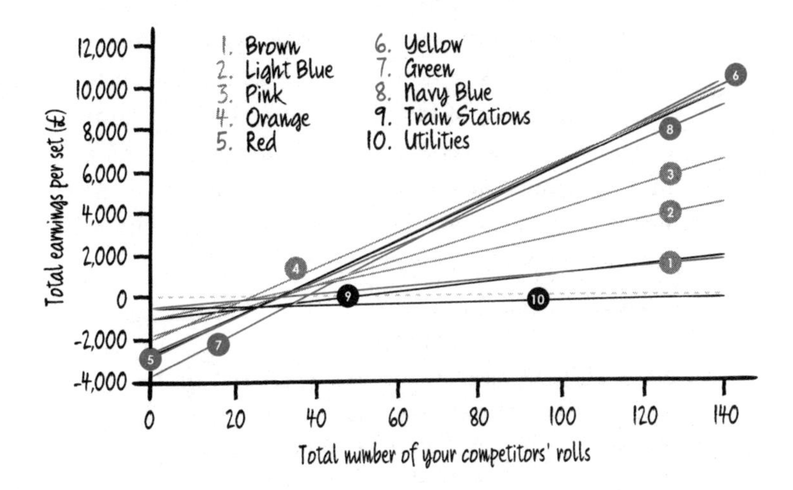

This graph gives a more accurate picture and tells quite a different story. Park Lane, which is very expensive to buy and build on but is rarely visited, ends up dragging down the Mayfair effect. The navy-blue set also has only two rather than three squares. This does mean it requires a smaller investment than the greens and so starts higher on the y axis, but it also means fewer opportunities to pull in the cash, hence the shallower slope. And so, dramatic as it might be when someone lands on Mayfair, it's just too rare an event to be worth the backing.

There is no single set that holds the key to winning the game, but there is always one set which will be the best for you to target. And the secret to which lies in the x axis.

At all times, you are looking to hold the property whose line is at the top of the pile, which changes depending on how long the game goes on. And since the average game of Monopoly takes about 30 turns per competitor, the set you want will change depending on how many opponents you have. More opponents means more turns, and hence it makes more sense to put your money into longer-term investments.

TOP TIP 2
When playing against just one opponent, go for the orange or light blue sets (both if you can). The same is true in a game with more opponents, but only in the early stages.

TOP TIP 3
In a game against two or three opponents which is likely to go on for a while, it's orange and red that you want to be targeting.

TOP TIP 4
Any more than three opponents, and green becomes your best shot for a chance at success.

It's also worth investigating the benefits of house buying and whether hotels really are the best buildings to aim for. Eventually any investment in Monopoly will make its money back, but early on in the game should you invest in building or hang on to the cash in case you find yourself on the receiving end of a big bill? You can compare these two scenarios and look at how many rolls of the dice by your competitors it will take before you're better off investing in that extra house. Here are the results for the best performing property of each set when deciding to build 1, 2, 3 or 4 houses or a hotel. The lower the line on the y axis, the more quickly your property starts making you money:

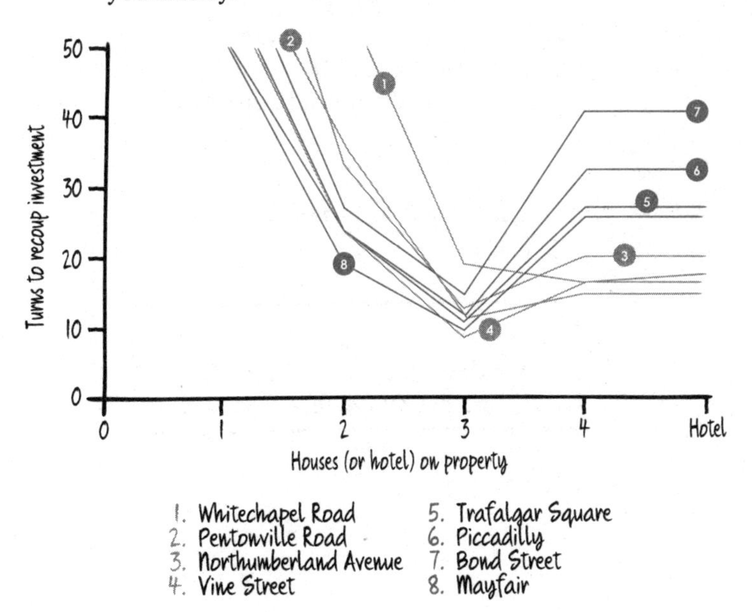

The first house takes a really long time to start paying for itself. For every property (apart from the brown set – which, let's be honest, is basically pointless) it's the third house that is really worth investing in quickly. After that, build more if you have the money, but it's probably worth waiting a few turns if cash is a bit tight. And since there are only a limited number of green plastic houses in the game, building three houses on properties early and then waiting to upgrade further has the added advantage of potentially blocking the building projects of other players. Sneaky, huh?

PRO TIP

Build three houses on your properties as quickly as you can and then wait until you have the cash to invest further.

But finally, if you take nothing else away from this entire chapter:

TOP TIP 5

Whatever way you look at it, utilities are completely pointless.

ENDNOTES

[1] There are a few ways to tackle this one, but in the interests of finishing the game before New Year's Eve, we'll just assume that at any point you have a 1/36 chance of throwing two consecutive doubles. The exact probability that you'd arrive at any particular square having thrown two doubles in a row is actually quite difficult to calculate because it depends on where you came from. 1/36 is a good approximation though, and isn't likely to change the results by more than around 0.001%. It's not perfect. So sue us.

Every even roll of the dice you then make will carry its own small risk of being your third double and hence sending you to Jail. All 2s you throw will have a 1 in 36 chance of sending you to Jail, since a 2 can only occur via a double. If you throw a four, there's a 1/3 chance it came from a double and so you'll have a $1/(3 \times 36)$ chance of going to Jail. And so on.

[2] We start with an imaginary player, say the boot (which is clearly the best Monopoly piece), and send it on a game around the board. But, rather than sending the boot to a single square on each turn, instead we will split it up into fractions of boot according to the probabilities of its arriving at each destination.

Starting the boot at Go, we get a version of the diagram on p.128 that shows not just the probability of where the boot will end up, but what fraction of the boot will be sent to each of those squares: 9.4% of the boot will be gutted at having to pay Income Tax on the first go, but not as annoyed as the 1.2% of the boot that finds itself in Jail.

These fractions of the boot now serve as the starting positions for its next turn. Of the 13.9% of the boot that landed on Euston Road, 5.6% will now go to Pall Mall, 8.3% will go to the Electric Company and so on and so on. Every time the boot takes a new turn it gets split into smaller and smaller fractions and spreads further and further across the board, and all the probabilities of these movements are dictated by the Markov matrix.

If that sounds like it's impossible to keep track of, then you haven't accounted for the remarkable minds of centuries of badass mathematicians. Naturally we've worked out a clever way to do all those calculations for us, in this case using something called matrix multiplication.

Eventually, there will be a tiny bit of boot on every square. Mathematically interesting, but a sartorial disaster. The fractions of boot will settle down on each square: on a new turn anything that leaves will be replaced by what arrives. And the result tells us how well visited each square is on the board across the course of a game.*

* A similar approach with slightly different assumptions was used to study the American version of the game in R. Ash and R. Bishop, 'Monopoly as a Markov process', *Mathematics Magazine*, vol. 45, no. 1, pp. 26–9.

CHAPTER 11

Watching Santa's weight

And so, almost as soon as it's begun, Christmas is over.

One moment you're dozing peacefully through the twilight hours of Christmas evening, too weary to switch over from whatever long-dead sitcom they've exhumed to fill the schedules; the next it's Boxing Day morning and you're wondering about the best way to get pine needles out of a DVD player and what on Earth to do with all those empty wine bottles.

The decorations may stay up till the sixth of January, but soon enough all signs of Christmas will have been purged from the house and whatever goodwill to all men you managed to muster over the festive season will have melted away, replaced by the miserable drudgery of everyday existence.

One thing that will take a little longer to recover, though, is your waistline. The bathroom scales won't forget about all that turkey, roast potato, chocolate and Christmas pudding, no matter how much you might want them to.

If you think things are bad for you though, spare a thought for Santa. Imagine how his heart must sink each time he clambers out of a fireplace to be confronted by yet another mince pie. Not to eat it would be out of the question – it would be terribly bad manners – so he wolfs it down and gets on with stuffing stockings, telling himself it will be the last of the night, but knowing that with all of North America still to visit, his indulgence is far from over.

Weighing in at 118 kg (260 lb) with a height of just 170 cm (5 ft 7 in), Santa has a body mass index of 40.7, putting him well beyond the threshold of 30 that indicates obesity.* However, despite his unhealthy dietary habits, research suggests Santa may actually be slimming down rather than getting fatter. In 2014, a study of 20 years' worth of Christmas card images found that Santa had lost over 2 stone in the preceding decade.†

This effort to shed some pounds may have been motivated by a bruising 1993 case in which a San Francisco court was asked to rule on whether Santa should 'present children with a slimmer, healthier image'.‡ Although the court decided in his favour, the experience must have been a sobering one for Santa. For a man who defines himself in terms of the joy and happiness he brings to children around the world, being accused of endangering their health must have been tough to swallow. It's easy to see why he might

* Santa's stats are from the North American Aerospace Defense Command, which has been tracking his Christmas Eve activities since 1955 (noradsanta.org).

† The study – carried out by greetings card company Clintons – was a serious scientific endeavour, definitely not a frivolous publicity gimmick ('Why a slimmer Santa will be coming to visit this year', *Telegraph*, 20 November 2014).

‡ 'Jolly old court upholds St Nicholas', *Los Angeles Times*, 12 December 1993.

have decided it was time to tighten his belt a little.

But if Santa wants to fight that flab, just how difficult might it be for him? While he does undoubtedly have to get through a lot of mince pies, not to mention as many glasses of sherry, for a man of his age he is also astonishingly active, climbing up and down thousands of chimneys and driving his sleigh at incredible speeds through the air. How much does that exercise offset all the calories he consumes?

There are two things we need to know. How many calories does Santa consume while he is out making deliveries? And how many does he burn?*

We're going to have to make some pretty big assumptions here. First, mince pies and sherry are far from a universal choice of gift for Santa. Children in different countries have different ideas about what he likes best, ranging from rice pudding in Denmark to pints of Guinness in Ireland.† We can't take into account every possible regional variation, though – not without going completely mad, at any rate – so we'll assume one mince pie and one 50 ml glass of sherry per household and hope that's sufficiently representative of the treats Santa is likely to come across throughout the night.

Second, Santa doesn't visit all the children of the world. As a great believer in religious and cultural

* Although there are disagreements over whether calorie-counting is actually a sensible way to control your weight (C. Wilson, 'Fat lot of good', *New Scientist*, 7 June 2016), it still forms the basis of official dietary advice in many countries.

† J. Savedge, 'What do kids around the world leave out for Santa?', *Mother Nature Network*, 23 December 2014.

tolerance, he understands that parents in some areas prefer to make their own arrangements for offering gifts to their children, so he respectfully avoids those homes where Christmas is not celebrated. As a very, very rough estimate of the size of his customer base, we'll consider the number of Christian children in the world and hope that Santa's many followers of other faiths or no faith more or less balance out the Christian children that Santa doesn't visit.*

Since there are about 1.9 billion children in the world and 31.4% of the world's population is at least nominally Christian, we'll estimate that Santa has to visit around 600 million children. Given that women have an average of 2.4 children throughout their lifetimes, we'd expect the average household to have fewer children than that,† so let's assume an average of 2 children for each address on Santa's route. That makes for around 300 million homes that Santa has to visit.‡

A quick browse through some supermarket websites suggests that the average mince pie contains

* There are some regions where Santa delegates his present-giving duties to various friends and allies (such as the Three Kings or an old woman), or where gifts are delivered on some day other than Christmas Eve. Gerry Bowler's *The World Encyclopedia of Christmas* (2000) provides a good summary of different customs.

† This is partly owing to child mortality and parental separation, but also to less tragic things, like the fact that women may not have completed their families yet or that some of their children may have grown up.

‡ All these statistics are from the US Central Intelligence Agency's *World Factbook* (cia.gov, 2016). We've defined children as anyone aged 0–14, because that's the age bracket they use (the next age bracket was 15–24). If you're aged 15–18 and feel you deserve a visit from Santa, don't blame us; blame the CIA.

about 250 calories, while 50 ml of sherry adds another 50 or so, for a total of 300 per home. That means that Santa's total calorific intake for the evening will clock in at around 90 billion, about 36 million times his daily requirement.*

If estimating the number of calories Santa takes in required some heroic leaps of imagination, estimating the number of calories he uses is, if anything, even tougher. Frustratingly little research has been conducted into how many calories you can expect to burn while climbing chimneys or flying a sleigh at supersonic speeds.

What we do know is how long Santa has to complete his deliveries. If parents ensure their children are asleep between 10 p.m. and 6 a.m., he can stretch the night out to a full 32 hours, using the 24 hours that it takes for Christmas Eve night to sweep round the world plus eight hours of leeway.† We have no idea how long he spends at each address, so let's take a shot in the dark and guess that he spends half his available time travelling.

Despite the lack of specific data on sleigh riding, the disturbingly thorough *Compendium of Physical Activities* does contain an entry for dog sledding (pretty similar, right?),‡ which suggests that, for a man of

* Based on a daily requirement of 2,500 calories (Source: nhs.uk).
† T. Chivers, 'Father Christmas's Christmas Eve in figures', *Telegraph*, 22 December 2009.
‡ B. E. Ainsworth et al., '2011 Compendium of Physical Activities: a second update of codes and MET values', *Medicine and Science in Sports and Exercise*, 2011, vol. 43, no. 8, pp. 1575–81.

Santa's weight, 16 hours of activity would burn 14,000 calories. Admittedly, the researchers probably weren't thinking in terms of supersonic huskies, but since the animals do most of the work anyway, we'll assume that the speed difference doesn't matter too much to Santa and we'll accept this as a fair estimate of his total in-flight energy expenditure.

As for those chimneys, at an absolute minimum Santa needs to expend 3 calories for every 10 metres he climbs, just to lift his own body weight.* His total energy expenditure will be much greater than that, though, since it's impossible to be 100% efficient, particularly when you have a stomach 'like a bowl full of jelly'.†

If we equate chimneys with 'high difficulty' rock climbing, returning to the *Compendium of Physical Activities* we find that a minute of this activity should cost Santa 15 calories. Distance is more important than time here, though, so if we suppose that a skilled rock climber ascends at a rate of 5 m per minute,‡ our best guess is that Santa should expend 3 calories per metre of chimney.

Taking an average of 5 m of chimney per

* To climb up a chimney, Santa needs to gain potential energy equal to his mass (118 kg) times the acceleration due to gravity (9.8 m/s^2) times the height gain required (10 m). The result is the required energy in Joules (11,564 J), which can then be converted into calories, giving a result of 2.8. Rounding up to 3 accounts for the weight of his clothes and an empty sack.

† C. C. Moore, 'A Visit from St Nicholas' (a.k.a. 'The Night Before Christmas'), 1823.

‡ V. Billat et al., 'Energy specificity of rock climbing and aerobic capacity in competitive sport rock climbers', *Journal of Sports Medicine and Physical Fitness*, 1995, vol. 35, no. 1, pp. 20–24.

household,* Santa has a total of 1.5 million km of climbing to do, expending a total of 4.5 billion calories.

This number is so huge that those 14,000 calories of sleigh riding pale into insignificance – but it is still only 5% of the 90 billion calories he's swallowed in sherry and mince pies! Even if we were to add in a few thousand calories for filling stockings,† and some more for going down the chimneys (surely easier than getting back up them), it's difficult to see how Santa could possibly use all that energy by the end of the evening. Given that those excess calories are equivalent to about 11 million kg of fat,‡ that weight loss is looking more impressive all the time.

Perhaps Santa's unique metabolism can be explained by looking to the animal kingdom. It doesn't seem much of a stretch to suggest that he might spend most of the year in hibernation.§ The behaviour fits perfectly: a burst of intense overeating followed by many months of inactivity. Granted, animals tend to

* This is probably a generous estimate. The typical height of a chimney for a two-storey house is a little under 10 m, but many children live in flats or houses without a chimney. However, since Santa always parks on the roof, he still has to climb back up one way or another.

† Perhaps equivalent to packing a suitcase, which the *Compendium of Physical Activities* reckons should cost Santa about 240 calories an hour, albeit at normal speed.

‡ 1 lb of body fat contains about 3,500 calories (M. Melnick, '3,500 calories equal a pound of fat?', *Huffington Post*, 5 March 2013).

§ There may be the odd clever clogs who tries to tell you that Santa's behaviour should actually be called aestivation, since he's sleeping through the summer rather than the winter. They're right, but we don't want to encourage pedantry, so try to pretend you didn't hear them.

reduce their energy consumption during hibernation while Santa will have to increase his by about 10 million per cent to use up his energy surplus,* but this still seems like the most reasonable explanation.

Come 26 December, then, Santa will already be tucked up tight in bed, alarm clock set for 1 December, red hat and coat hanging neatly from Rudolph's antlers, with an empty packet of super-strength indigestion tablets lying on the bedside table next to him.

And that's where we'll have to leave him. Unlike Santa, mathematicians never sleep, so we'll be heading back to our workshop to sit out the rest of the winter enjoying the simple pleasures of triangulating matrices, rationalizing denominators and integrating polynomials. Try not to be too jealous.

We hope you've enjoyed our mathematical meanderings through the rituals of Christmas. If you do try any of our top tips, be aware that we categorically refuse to accept any responsibility for the broken friendships, torn wrapping paper and general misery that might ensue. Obviously, though, if anything actually does work out, it was definitely down to us.

While it's true that throughout these pages we may have been guilty of the odd bit of cynicism, like most of the *Bah! Humbug* crowd we love Christmas really. And if there's one thing that Christmas and

* Assuming he wakes up on 1 December to give himself a few weeks to get ready and that he celebrates Christmas Day with the elves, Santa needs to use up 85 billion calories in the 340 days from 26 December to 30 November. That's 250 million calories a day. An average adult male would expect to be using about 2,500.

mathematics have in common, it's that neither of them can really be appreciated without keeping hold of a bit of curiosity, imagination and good old-fashioned childlike wonder.

So after all those graphs and equations, there's only one thing you really need to take away with you. If you've been concentrating – and you haven't just turned to this page to find out how the book ends – you'll already know what it is.

Whatever you do, there's one thing you must never ever forget . . .

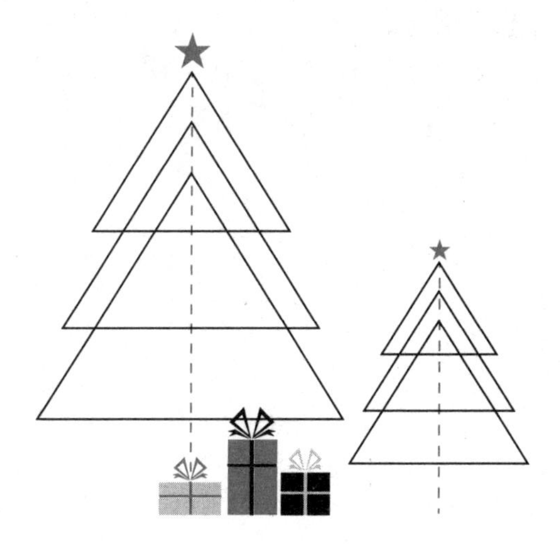

SANTA
CLAUS
exists . . .

ACKNOWLEDGEMENTS

Our little book had some big help from a number of wonderful people: Susanna Wadeson and the whole team at Transworld, Claire Conrad from Janklow & Nesbit, Gillian Somerscales, Julia Lloyd, Lis Adlington and Anna Gregson. Thanks too to Andy Hudson-Smith and the team at CASA for their continued support.

Hannah: Thanks to Tracy, Natalie, Marge & Parge, Mr P, Miss McGee and little Edie. I'm very lucky indeed to have you on my team.

Thomas: Thanks to my fantastic family – Emily, Gareth and Catherine Evans – for all manner of reasons, and to the wonderful Emilie Oléron Evans, for very graciously sharing me with Santa over the past few months. Thanks also to Dennis Barnett and Elizabeth Swan, two tremendous maths teachers, who had a very significant influence on my life and the lives of so many others. Finally, back in 1996, I promised to dedicate my first book to my uncle. I'm finally able to make good on that promise, so my share of this text is dedicated to Michael Evans; thanks for your imagination and for all those Target novelizations.

Dr Hannah Fry is a mathematician from University College London. In her day job she uses mathematical models to study patterns in human behaviour, from riots and terrorism to trade and shopping. You'll also find her on BBC radio and television, where she regularly presents science programmes, or on YouTube's Numberphile channel.

Photograph by Emilie Oléron Evans

Dr Thomas Oléron Evans describes himself as a mathematician and a writer, though how others would describe him is anyone's guess. He used to be a teacher, but managed to escape and now has a research and lecturing job in London. While broadly agreeable, this does make life a little complicated, since he lives in Strasbourg, France.

For more information about the authors, visit their websites, www.**hannahfry**.co.uk and www.**mathistopheles**.co.uk, or follow them on Twitter **@FryRsquared** and **@mathistopheles**.